BIOCHEMISTRY RESEARCH TRENDS

X-RAY OPTICS AND INNER-SHELL ELECTRONICS OF HEXAGONAL BN

BIOCHEMISTRY RESEARCH TRENDS

Additional books in this series can be found on Nova's website under the Series tab.

BIOCHEMISTRY RESEARCH TRENDS

X-RAY OPTICS AND INNER-SHELL ELECTRONICS OF HEXAGONAL BN

ELENA O. FILATOVA
AND
ANDREY A. PAVLYCHEV

Nova Science Publishers, Inc.
New York

Copyright © 2011 by Nova Science Publishers, Inc.

All rights reserved. No part of this book may be reproduced, stored in a retrieval system or transmitted in any form or by any means: electronic, electrostatic, magnetic, tape, mechanical photocopying, recording or otherwise without the written permission of the Publisher.

For permission to use material from this book please contact us:
Telephone 631-231-7269; Fax 631-231-8175
Web Site: http://www.novapublishers.com

NOTICE TO THE READER

The Publisher has taken reasonable care in the preparation of this book, but makes no expressed or implied warranty of any kind and assumes no responsibility for any errors or omissions. No liability is assumed for incidental or consequential damages in connection with or arising out of information contained in this book. The Publisher shall not be liable for any special, consequential, or exemplary damages resulting, in whole or in part, from the readers' use of, or reliance upon, this material. Any parts of this book based on government reports are so indicated and copyright is claimed for those parts to the extent applicable to compilations of such works.

Independent verification should be sought for any data, advice or recommendations contained in this book. In addition, no responsibility is assumed by the publisher for any injury and/or damage to persons or property arising from any methods, products, instructions, ideas or otherwise contained in this publication.

This publication is designed to provide accurate and authoritative information with regard to the subject matter covered herein. It is sold with the clear understanding that the Publisher is not engaged in rendering legal or any other professional services. If legal or any other expert assistance is required, the services of a competent person should be sought. FROM A DECLARATION OF PARTICIPANTS JOINTLY ADOPTED BY A COMMITTEE OF THE AMERICAN BAR ASSOCIATION AND A COMMITTEE OF PUBLISHERS.

Additional color graphics may be available in the e-book version of this book.

LIBRARY OF CONGRESS CATALOGING-IN-PUBLICATION DATA

Filatova, E. O. (Elena Olegovna)
  X-ray optics and inner-shell electronics of hexagonal BN / Elena O. Filatova and Andrey A. Pavlychev.
       p. ; cm.
   Includes bibliographical references and index.
   ISBN 978-1-61209-260-7 (softcover)
   1. Boron compounds. 2. Crystallization. 3. X-ray optics. I. Pavlychev, A. A. (Andrei Alekseevich) II. Title.
   [DNLM: 1. Boron Compounds--chemistry. 2. Crystallization--methods. 3. Optics and Photonics. 4. X-Rays. QD 181.B1]
   QD181.B1F55 2011
   546'.6712--dc22
                              2011004001

Published by Nova Science Publishers, Inc. † New York

CONTENTS

Preface		vii
Introduction		1
Chapter 1	Crystal Structure and Chemical Bonding of Hexagonal BN	3
Chapter 2	Band Structure and Density of States for Hexagonal BN	5
Chapter 3	Quasi-atomic Approach to X-ray Absorption	17
Chapter 4	Boron and Nitrogene Excitations	37
Chapter 5	Optics of Hexagonal BN	51
Chapter 6	X-ray Radiation Interaction with Hexagonal BN Crystal	59
Chapter 7	B and N K - Absorption Spectra of h-BN	73
Chapter 8	X-Ray Polaroid	87
References		93
Index		103

PREFACE

In the chapter special emphasis is put on interaction of x-ray polarized and non-polarized radiation with hexagonal BN (h-BN) crystal. h-BN is one of the most anisotropic layer compounds. sp^2 hybridization controlling chemical bonding of boron and nitrogen atoms causes strong anisotropy of the electronic and atomic properties along and perpendicular to the atomic layers in the crystal. Near the B and N K shell ionization thresholds the dielectric tensor in h-BN demonstrates sharp variations with photon energy and results in intriguing optical properties of the compound. X-ray absorption near the thresholds is strongly influenced by the short range order parameters of the layers and demonstrates the pronounced "molecular" properties. The following chemical, structural and optical aspects of h-BN are discussed in more detail: 1) crystal structure and chemical bonding; 2) ground-state properties, band structure and density of states; 3) quasi-atomic properties of x-ray absorption and inner-shell photoemission; 4) x-ray optics of h-BN; 5) orientation and polarization effects in the x-ray reflection, photon scattering and absorption; 6) x-ray Polaroid.

h-BN is selected by the authors as the basic object of the investigation for it demonstrates the most characteristic behavior in X-ray optics and inner-shell electronics. This phenomenon is considered by the authors as a prototype for numerous trends in biochemistry research.

The authors acknowledge numerous and fruitful discussions with Prof. A. S. Shulakov and Prof. A. S. Vinogradov. The authors also thank Dr.Taracheva and Dr.Sokolov for providing results to publication.

INTRODUCTION

Extensive work during the past two decades on the optical and electronic properties of boron nitride and carbon allotropes has outcome a lot of significant experimental and theoretical results and led to many interesting hypothesis concerning their link with the short and long range order parameters. Here special emphasis is put on interaction of x-ray radiation with hexagonal boron nitride (h-BN) and on linkage of the spectroscopic data with the electronic, atomic, molecular and optical properties of the solid. X-ray absorption and emission spectra are conventionally regarded as a promising probe of local electronic and atomic properties and chemical bonding in ground and core-ionized matter. Measurements of x-ray reflection, transmission and refraction are the sensitive tool of the determination of the optical constant of media. Cooperative research in x-ray optics and inner-shell electronics opens a way for more detailed understanding of the interplay of the short and long range order parameters and the optical constants as well as their linkage with the x-ray spectra.

X-ray absorption and inner-shell photoemission processes are traditionally described by using the band structure theory of a solid or the quantum chemistry methods elaborated for clusters. However they do not entirely help with the specifics of x-ray absorption involving the formation of highly spatially localized core-excitations in solids. In contrast, the quasiatomic approach to x-ray absorption is widely used here to take effectively into account the dynamic core-hole localization in equivalent atoms in solids. Reflection, transmission and refraction of x-ray radiation are traditionally described by using the Fresnel's equations. To bring together x-ray absorption and reflection spectroscopic results the integral Kramers – Kronig dispersion relation is applied. Joining these approaches allows a broad outlook on electronic, atomic, optical and chemical properties of h-BN.

Boron nitride [1] is binary compound of a boron and nitrogen and can exist both in crystal and amorphous modifications as well as in a form of nanotubes.

Crystal boron nitride is similar to carbon and it also exists in several allotropic modifications: hexagonal (α), similar to graphite; cubic (β) blende type, similar to diamond; (borazon) and dense hexagonal (w) wurzite type.

The greatest interest for the modern techniques represents hexagonal boron nitride (α-BN or h-BN). Graphite and hexagonal boron nitride are among the most popular layered materials used for the interface engineering. Although h-BN is isostructural and isoelectronic to graphite, the partial ionicity of the B-N bonds pushes the highest occupied and lowest unoccupied states apart making h-BN an insulator with the fundamental band gap of ~ 3.0 - 7.5 eV [2-29]. It is more stable than graphite in high pressure–temperature machining [30]. Hexagonal BN has a melting point of >3000 K [31], and is an excellent thermal conductor. Its high melting temperature and low coefficient of thermal expansion make it useful in vacuum technology, and as crucible coatings [32] for holding molten metals. Hexagonal BN is also the basis for BN nanotubes [33] and it is raw material for cubic BN (borazon) fabrication which hardness is the same as at diamond [34]. The ease of sliding between basal planes makes it a great solid lubricant [35] for reducing wear and friction, and it can be added to other solid/liquid lubricants for machining processes [36].

H-BN can be produced by several methods: chemical reactions between boron compounds and nitrogen compounds [37] (e.g. boric acid + urea), hot pressing, combustion synthesis [38], and chemical vapour deposition [39].

Chapter 1

CRYSTAL STRUCTURE AND CHEMICAL BONDING OF HEXAGONAL BN

Hexagonal BN is one of the most anisotropic layer compounds. In ordinary conditions h-BN crystal is composed of graphitelike layers located perpendicular to main crystallographic axis of crystal (direction [0001]), which is also optical axis of the crystal. In each atomic layer (basal plane) three atoms of N surround each atom of B and vice versa (Figure 1). Concerning to interlayer structure there are many ways (stackings) in which one plane can be placed above the adjacent plane [29]. The point and space symmetry group of h-BN follows the symmetry of different stackings. Results from the calculations [29] showed that only two configurations are stable in that the displaced plane moves back to its original position. They belong to D_{3d} and C_{3v} point symmetry group or D_{6h} and D_{3h} point group (according to Schoenflies). The theoretical mass density of h-BN is 2.271 g cm^{-3} [21]. The lattice constants are $a_0 = b_0 = 2.494$ Å, and $c_0 = 6.66$ Å [21], which are very similar to those of graphite: $a_0 = 2.5$ Å and $c_0 = 6.7$ Å. The parameter of anisotropy calculated as ratio c/a is equal to 2.67.

One can see that $2p_z$-atomic orbital doesn't participate in the hybridization process. Such chemical bonding leads to the origin within each layer of trigonal σ bonds with participation of three valence electrons from boron and three from nitrogen. This bonding is covalent mixed with a touch of ionic. The π orbitals are formed by the overlap of N2p_z orbitals containing two p electrons with empty B2p_z orbitals, which are perpendicular to layers. Interplanar bonding is very weak with no directional bonds [40] present, and is probably a mixture of ionic attraction between oppositely charged ions in adjacent planes, and van der Waals bonding such as in graphite.

The strong directional bonding between adjacent coplanar atoms shows charge localization closer to the N atom than the B atom, and depending on the radii assumed for each atom, each B atom loses 1–2 electrons to its three neighboring N atoms. Electrons in π orbitals are also localized closer to the N atoms than the B atoms [41].

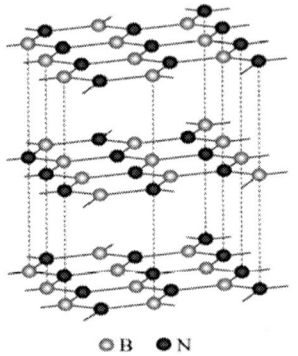

Figure 1. Crystal structure of hexagonal BN.

The chemical bonding of h-BN is commonly described by sp^2 hybridization of boron and nitrogen orbitals: $2s + 2p_x + 2p_y = 3(2sp^2)$. The process of sp^2 hybridization is shown on scheme (Figure 2):

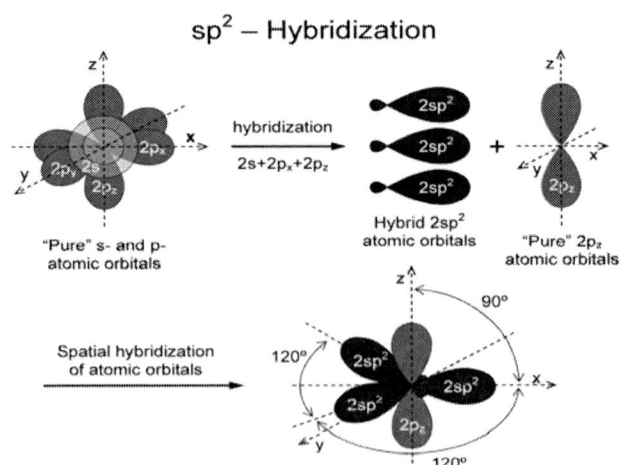

Figure 2. The process of sp^2 hybridization.

Chapter 2

BAND STRUCTURE AND DENSITY OF STATES FOR HEXAGONAL BN

One of the effective methods of the analysis of the electronic structure of solid states is X-ray spectroscopy. Near edge spectral dependencies of the x-ray absorption coefficient and spectral distributions of the intensity in the characteristic x-ray emission bands reflect the energy distribution of the density of empty electronic states of the conduction band (CV) and occupied electronic states of the valence band (VB), respectively. X - ray absorption and emission processes have a local character (associated with hole localization in the core shell) and dipole selection rules for the transitions between the initial and the final state have been worked out.

Figure 3 presents schematically X-ray absorption and emission processes. The former results in the pumping of photon energy in a solid and the ejection of an electron from a core (atomic) level to unoccupied states in the conduction band and the latter refers to electronic transitions from the valence band to the partly unoccupied core level and photon emission.

Thus the possibility to obtain the information about local and partial (allowed for certain angular momentum symmetry) density electronic states of the conduction and the valence band is appeared. Such unique information does not possess a single method.

The experimental results shown in figure 4 represent K (1s)-spectra of boron and nitrogen of h-BN studied by Fomichev [6].The energy positions of the main spectral features are given in tabl.1.

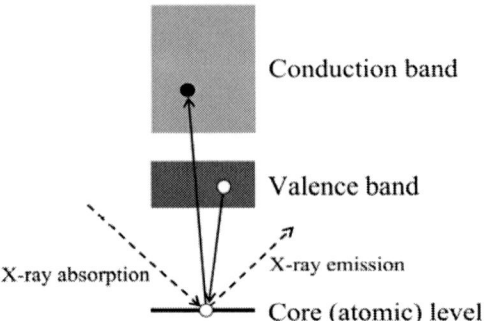

Figure 3. Schematic presentation of X-ray absorption and emission processes in a solid.

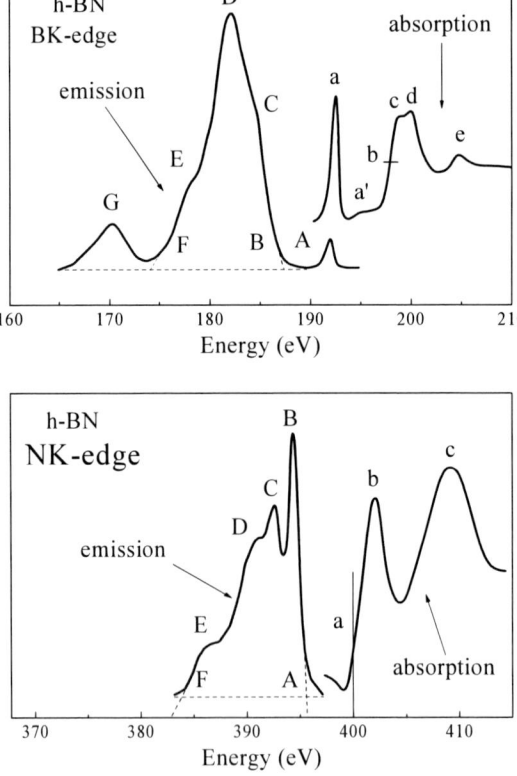

Figure 4. K (1s)-spectra of boron and nitrogen of h-BN, studied by Fomichev [6].

Table 1. Energy positions (eV) of the main spectral features in K-spectra of h-BN

Emission spectra	G	F	E	D	C	B	A
B K-	170.2	175.4	177.5	181.5	184.0	186.4	191.55
N K-	-	384.6	386.5	391.0	393.0	394.7	395.6

XANES spectra	a	a'	b	c	d	e
B K-	191.95	194	197.4	198.0	199.3	203.9
N K-	402.1	-	402.1	409.0	416.6	-

As follows from the figure 4 the widths of the main K-emission bands of boron and nitrogen estimated on the spacing B-F in the spectrum of boron and A-F in the spectrum of nitrogen are equal to 11.0±0.5 eV. At the same time there are substantial differences in BK- and NK-spectra of h-BN. A narrow selective maxima (A – in the emission spectrum and a – in the spectrum of quantum yield) are observed only in the spectra of boron. Moreover, the separation between points B (high-energy edge of the valence band) and b (K-absorption edge of boron in h-BN) is equal 11 eV. In contrast the separation between the end of the emission band (point A) and beginning of the absorption in the spectra of nitrogen is equal 4.5 eV. Such considerable distinction of the magnitude of gaps (between the end of the emission band and the beginning of the absorption) in the spectra of component of h-BN leads to a problem of correct coincident of the BK- and NK- spectra in the common energy scale.

The correct combination can be achieved with help of x-ray photoelectron spectroscopy (XPS) data. In the XPS method is assumed that there is thermodynamic equilibrium in the system "sample-spectrometer" so that Fermi levels of sample and spectrometer are equalized. Then if the binding energies E_c of electrons of sample are measured from common Fermi level the relation:

$$h\nu = E_c + \varphi + E_k \tag{1}$$

where $h\nu$ is energy of x-ray photon;
   φ is photoelectric work function;
   E_k is kinetic energy of photoelectron

is valid. The equation (1) allows calculating of binding energies E_c of all electrons participating in photoeffect basis on measured E_k. The diagram of x-ray photoelectron spectra is plotted in the figure 5a. As follows from figure 5a the magnitudes of binding energies, obtained on a basis of XPS data can be used to compare the x-ray spectra of different atoms constituting the crystal in the common energy scale.

The XPS spectra of h-BN obtained by Hamrim and co-authors [42] are plotted in the figure 5b.To compare the x-ray emission spectra in the common energy scale it is enough to shift these spectra in the energy scale on the value, which equals to the difference of energies of K-levels of boron and nitrogen. The following values of binding energy relatively Fermi level can be established: B(1s) – 190.6 eV, N(1s) – 398.3 eV, N(2s) – 19.4 eV.

BK and NK – spectra aligned in the common energy scale are presented in the figure 6. The Fermi level was chosen as zero-point of the energy scale (in the absolute energy scale of BK-spectrum it is equal 190.6 eV).

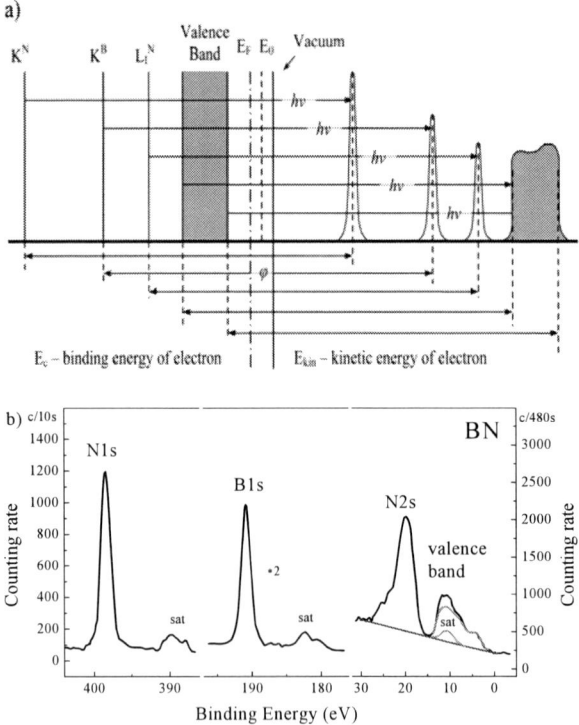

Figure 5. Diagram of x-ray photoelectron spectra (a); XPS-spectra of h-BN obtained by Hamrim and co-authors [42] (b).

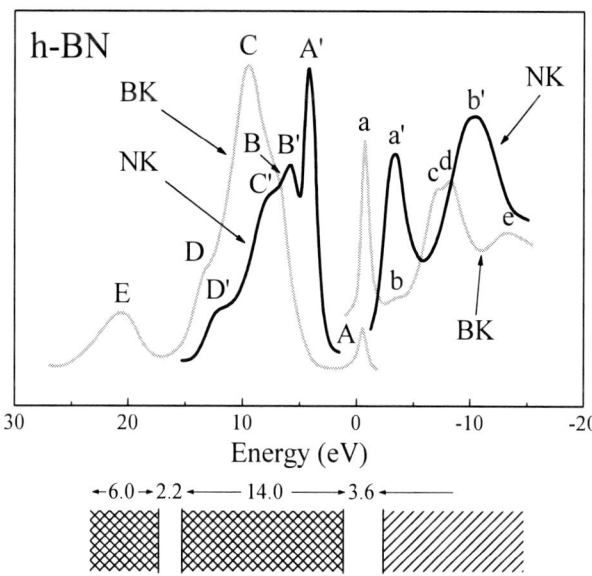

Figure 6. BK and NK – absorption and emission spectra of h-BN (figure 4) aligned in the common energy scale.

It is necessary to note that absolute values of core level energies of atoms in compounds measured by different authors using XPS method are noticeably differed. However the differences of the energies of these levels are much closed and combination of spectra can be spent to within 0.2 eV [42-44]. As follows from the figure, the distance from Fermi level to bottom of conduction band equals 1.8 eV. If the bottom of the conduction band is associated with the beginning of the absorption in NK spectrum, then the width of band gap is equal to 3.6 eV. The joint consideration of BK and NK spectra leads to the conclusion that the electronic states in the vicinity of the top of the valence band are reflected in the K-emission spectrum of nitrogen, only. Then, in the direction of the bottom of valence band there is a redistribution of the intensities of bands belongs to boron and nitrogen and near the bottom of valence band the electronic states mainly reflect the states of boron atom.

Let us now address to theoretical studies of h-BN. There have been many studies of the electronic properties of h-BN. Early studies show vastly different results mainly because of differences in the computational methods.

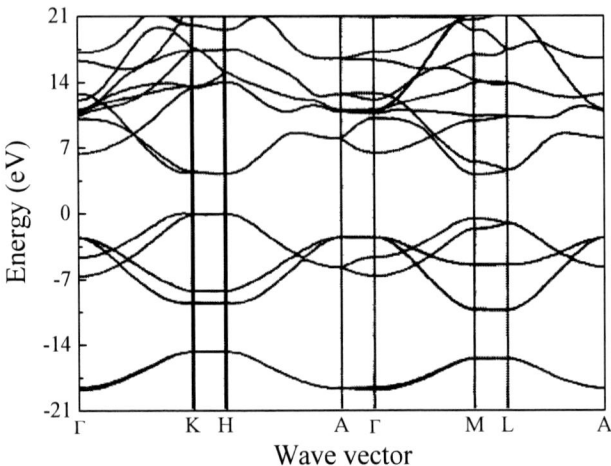

Figure 7. Calculated band structure of hexagonal BN [21].

The band structure for h-BN calculated self-consistently using the OLCAO (orthogonallised linear combination of atomic orbitals) method by Xu and Ching [21] is plotted in the figure 7. The grahpitelike layer structure of h-BN is reflected in the relatively flat bands along the k_z direction. The calculated band gap is indirect and has values of 4.07 eV that agrees reasonably with the experimental data obtained by Fomichev [6]. For h-BN, the direct and indirect band gaps are quite close. As can be expected, the effective mass (EM) for h-BN are highly anisotropic with average in plane components a factor 5 smaller than the components perpendicular to the plane [21]. According to calculations by Xu and Ching [21] along the Γ to M direction, the electron and hole EM components are 0.26 and -0.50, respectively, thus the charge carriers in the basal plane of h-BN can be quite mobile.

VALENCE BAND

The early calculations of the density of states (DOS) of the valence band of h-BN have been carried out for a two-dimensional unit cell (2 atoms – one boron and one nitrogen atom) using tight binding approximation [8, 45], three-dimensional unit cell (4 atoms – two boron and two nitrogen atoms) or set of parallel layer planes using tight binding approximation [19], or orthogonalized plane wave method [11]. For the expanded cell it was used the OLCAO [17, 21]. The calculations realized by different methods in the framework of different models have shown similar structure of the valence band of h-BN. On the figure 8 (a) the results of DOS calculation for three-dimensional unit cell carried out by OPW method [11] are presented. As follows from DOS distribution, three main regions can be distinguished in the valence band of h-BN: the π-band at low binding energies overlapping with the σ-band, which is separated by an energy gap from so-called s-band. The σ-band has mixed s+p character and reflects intra-plane bonding. The π-band consists exclusively of p-states and reflects inter-plane bonding in h-BN.

The partial density of states (PDOS) of boron and nitrogen atoms calculated for three-dimensional unit cells using density functional theory (DFT) by Ooi and co-authors [29] is plotted in the figure 8 (b, c). As follows from the figure, there is a substantial difference in the PDOS of boron and nitrogen atoms. It is obviously, total DOS distribution of h-BN to a greater extent coincides with the PDOS distribution for nitrogen that means the states of the valence band are essentially more spatially localized around the nitrogen atoms than around the boron atoms. Comparison of DOS and PDOS distributions obviously shows that the s-band mainly originates from 2s states of nitrogen; the π – band contains only p_z - electrons of nitrogen and σ – band connects with p_x, p_y – components of both boron and nitrogen and contains small impurity of s-like states of boron, which are concentrated mainly at the bottom of valence band.

The calculations carried out by different authors in the framework of different models have demonstrated the similar structure of the valence band of h-BN. In contrast different calculations give very diverging widths for some bands; for example, the computed widths of the crucial π-bands differ by a factor of 4 [8, 10, 11, 17]. Robertson [19] have found that π-bands are anomalously wide because the π interaction between two p_z orbitals is almost twice that between two $p_{x,y}$ orbitals, indicating the σ and π orbitals to have very different spatial extents. That means the width of the valence band strongly depends on π – interaction between two p_z orbitals, which does not taking into account in two-dimensional model.

Nakhmanson and Smirnov [10] have calculated the valence band structure in three-dimensional model by orthogonalized plane wave method. The calculations were carried out in all points of high symmetry and for all symmetric directions (the basis of calculation included 73 orthogonalized plane waves). Also the same calculations have done for the basis of calculation included 220 orthogonalized plane waves. It has established that taking into account three-dimensional structure leads to doubling of bands in contrast to calculation for a two-dimensional carried out by the tight binding approximation. The calculated width of valence band has made from 8 to 30 eV by different authors and strongly depends on model and method of the calculation [8, 10, 11, 19, 29, 45, 46].

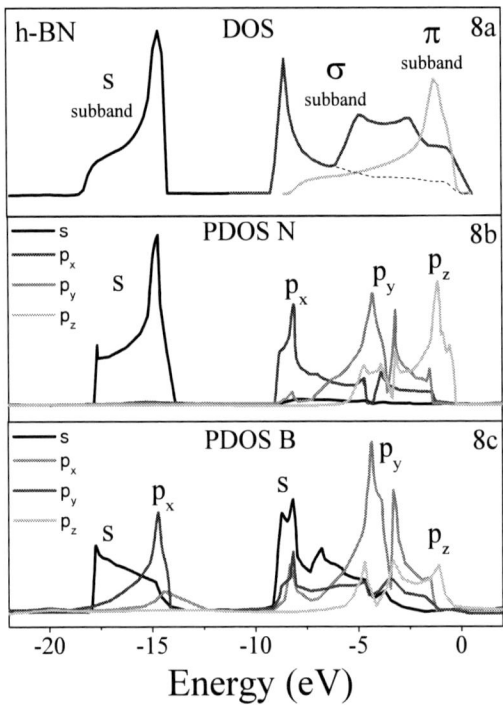

Figure 8. Calculations of DOS and PDOS for h-BN taken from the works [11] and [29].

The experimental studies of the valence band of h-BN using x-ray emission spectroscopy one can find in the works [6, 12, 53]. As was mentioned above the BK- and NK-emission bands are closely related to the local partial p-density of states in the environment of the B and N atoms, respectively and allows to study in details the structure of valence band.

For a comparison of experimental results with the π- and σ-density states calculated by Ooi and co-authors [29], the BK- and NK- spectra of h-BN (figure 4) and PDOS calculations of boron and nitrogen [29] are aligned in common energy scale in the figure 9. The peak of theoretical $\pi(p_z)$ -band was aligned with the sharp peak of the π-band of the nitrogen emission spectrum. One can see it is reasonable agreement between the experimental data and the corresponding structures in the calculated PDOS. One can see that p_z- states connected with π-electrons and in great part localized on the nitrogen atoms is sharply defined in NK-emission spectrum (very narrow band B) and doesn't pronounced in BK- emission spectrum. C, D, E details of the structure in BK emission spectrum and C', D', E' in NK-emission spectrum reflects a mix of p_x+p_y states of boron and nitrogen, respectively, connected with σ-electrons. Detail G in BK-emission spectrum essentially reflects p_x states of boron. According to PDOS of boron the s-states, mixed with p_x states, should be reflected in the BK –emission spectrum. Nevertheless because of pronounced Auger-broadening of the spectra a comparison between theory and experiment for the s-band hardly is possible.

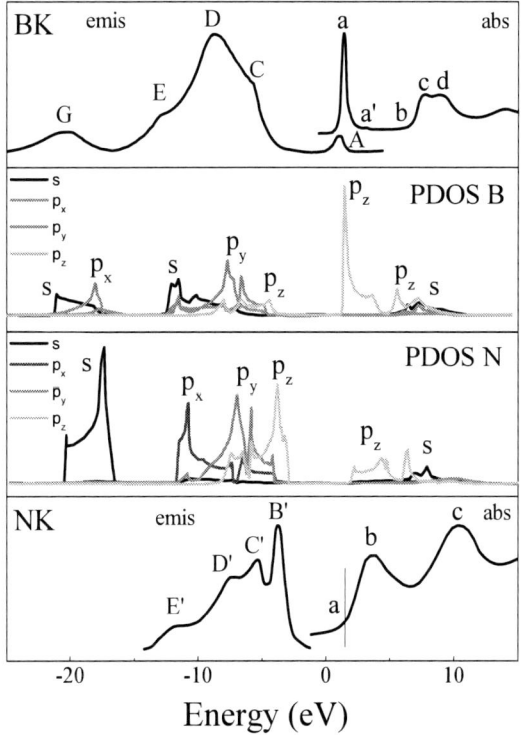

Figure 9. BK- and NK- spectra of h-BN (figure 4) and PDOS calculations of boron and nitrogen [29] aligned in common energy scale. The contributions of different partial waves centered at boron and nitrogen atoms to the PDPOS are also shown.

The width of the valence band obtained from PDOS distributions is equal about 20 eV that agree reasonably with the spectroscopic results. This value is in a good agreement with the width derived from XPS data, too.

BAND GAP

Because of the great discrepancies between values of band gap obtained by different authors using different experimental methods and theoretical models the width of band gap of h-BN is a hot topic up to now.

Table 2. Experimental values of band gap energy for h-BN

Method	E_g (eV)
X-ray emission spectra	3.6 [6]
X-ray emission spectra	3.85 ± 0.5 [12]
Optical absorption spectra	3.9 [7]
UV absorption spectra	4.3 [13]
Reflection spectra	4.5 [9]
Optical reflectivity	5.2 ± 0.2 [20]
Multiband luminescence spectra	5.5 [4]
Optical spectra	5.89 [22]
Reflection, absorption and photo-conductivity spectra	5.8 [15]
Absorption spectra	5.83 [14]
Temperature dependence of electrical resistivity	7,1 [18]
nanocrystalline films absorption spectra	6.0 [51]
inelastic electron scattering	5.9 [52]

The early calculations carried out for two-dimensional unit cell [8] using tight-binding approximation give the band gap of 5.4eV. A simple empirical analysis for the prediction of the energy gaps in III-V and in (rare earth)-V semiconductors were developed by Sclar (1963). This analysis is based on a correlation of energy gaps with the ionic and covalent atomic radii of the constituent elements. The value 7.25 eV was predicted for band gap of h-BN. The orthogonalized plane wave method gives a value 3.8 eV. Ooi with co-authors [29] have calculated the band structure of h-BN using DFT method for two-dimensional and three-dimensional unit cell and received the value 3.6 eV.

The experimentally obtained values of band gap for h-BN are summarized in the table 2. As follows from the table there is a wide scatter of the value of band gap (from 3.0 to 7.5 eV) derived from different experiments. The established discrepancies between band gaps derived experimentally can be explained by different technologies of manufacturing of investigated h-BN samples. Depending on technology one can prepare mono or poly crystalline [18, 28] or thin film of nanometers thickness [47, 48, 51] or powder [49, 50]. Besides, exactly the technology of manufacturing of h-BN defines the quantity and chemical composition of impurity in the sample, its porosity and symmetry of stackings. The value of band gap depends on these factors.

CONDUCTION BAND

The conduction band of h-BN have calculated using different methods based on the band theory of a solid, the MO LCAO approach and the multiple scattering, The results of calculations realized by different methods have shown similar band structure of h-BN. The PDOS calculations of the conduction band for boron and nitrogen carried out using DFT calculations [29] is presented in the figure 9. As can be seen the conduction band consists of narrow π-subband, arranged in the vicinity of bottom of conduction band, and σ-subband. According to the PDOS calculations the p_z- states connected with π-electrons is sharply defined in BK-absorption spectrum and only slightly pronounced in NK- absorption spectrum. The σ- subband represents a mix of $s+p_x+p_y+p_z$ components. The intensity of π- subband in PDOS of boron is considerably superior to intensity of σ-subband. This conclusion agrees well with experimental BK- absorption spectrum [6, 59].

The well elaborated methods based on the band theory of a solid, the MO LCAO approach and the multiple scattering, however, do not entirely allow for the specifics of radiation absorption of the core shells of the atoms, involving the formation of highly spatially localized excited states possessing high sensitivity to short-range order in solids. It is extremely difficult in terms of the enumerated theoretical methods to analyze the propagation of electrons produced by the absorption of an X-ray quantum in complex polyatomic systems. It obviously the dynamic effects that determine for the most part the structure of the localized excited states of a solid and the spectral features of the electronic and optical phenomena, as well as the quality of the short-range order parameters should to be considered.

Chapter 3

QUASI-ATOMIC APPROACH TO X-RAY ABSORPTION

BASIC RELATIONSHIPS

The symmetry dependence of unoccupied states and their link with atomic partial waves in h-BN chemical compounds can be understood in the framework of the quasi-atomic (QA) approach [54-61], which allows one to describe X-ray absorption fine structure in terms of strongly localized atom-like orbitals in polyatomic systems. Let us introduce an atom-in-a-compound effective potential $V^{(\lambda)}(r) = v(r) + W^{(\lambda)}(r)$ depending on the parameter λ, characterizing the influence from the surrounding atoms on the core-excited atom in the solid [54,57]. v is the potential of core-excited atom (or ion). When $\lambda \to 0$ the surroundings potential $W^{(\lambda)}$ comes to nil. In particular, the ratio R_e/R of equilibrium R_e to variable R interatomic distances can be taken as the parameter. Evidently that in this case $V^{(\lambda)}(r)$ centered at the core-excited atom varies from the potential $v(r)$ to the potential $V(r)$ of the atom (ion)-in-compound under study, when $\lambda = 1$, and to united atom, when $\lambda \to \infty$. In general case $W^{(\lambda)}$ is a nonlocal and energy dependent pseudopotential [54]. The methods of its construction are described in the works [54, 62-64]. The effective and surrounding potentials possess the point symmetry group of a core-excited atom in a compound. The symmetric aspects of core excitations are tightly related with the dynamic core-hole localization.

Neglecting by weak van-der-Waals interaction between atomic layers in h-BN we approximate the effective potential for the atomic-like np_z ($np\pi$) states oriented perpendicular to the atomic layers as $V_\pi^{(\lambda)} \approx v$. In many cases this is reasonable approximation for $0 < \lambda \leq 1$ [54, 55]. In contrast, when $\lambda \to 1$ the effective potential $V_\sigma^{(\lambda)}$ for the atomic $np_{x,y}$ ($np\sigma$) states oriented along the layers differs substantially from the atomic potential. As an illustration taken from the works [54, 55], figure 10 presents what happens with 1s → $2p_x\sigma$ transition in the core-excited nitrogen atom in the model N^+ --- N system oriented along the axis x when the parameter λ increases from 0 up to 1.2. For λ close to unity the effective potential is essentially inhomogeneous and has the specific two-valley shape (see, figure 10, upper plot). These valleys are separated by the potential barrier. The potential (pseudopotential) barrier is originated by the combination of the centrifugal repulsion and the screening of the attractive potential of core-ionized atom by the neighboring atoms.

When λ increases the effective potential causes the blue shift of the N 1s → $2p\sigma$ resonance in X-ray absorption from the discrete part to the continuum and the broadening of the resonance in the continuum. This behavior of the atomic resonance is associated with the extraction of the atomic resonance from the inner well when the interatomic distance R is shortened. Changes in chemical bonding in the model system are not taken into account. Although the resonance cannot be described in the continuum as a 2p state their familiar relation is seen clear and allow us assigning the resonance with the σ (2p) shape resonance. Similar changes in spectral distribution of oscillator strength appear in planar compounds, in particular, in electronic transitions from B 1s shell in the BF_3 molecule. The changes referring to the variation of interatomic B – F separation, are presented in the work [54].

Thus, in linear and planar systems one may speak about the splitting of the triply degenerated atomic 1s → 2p resonance into two components the lowest (π) and highest (σ) of them are respectively assigned with atom-like transitions from deep 1s level to doubly (or singly in planar systems) degenerated $2p\pi$ states oriented perpendicular to the directions on neighboring atoms or singly (or doubly in planar systems) degenerated σ(2p) states corresponding to σ-bonds in the compounds. This quasi-atomic treatment is supported by the investigations of K-shell excitations in linear diatomic and triatomic molecules such as N_2, CO, CO_2, N_2O, planar molecules BF_3, BCl_3, BBr_3, C_6H_6, as well as planar h-BN and different cubic systems such as SF_6, LiF, NaF. We stress that the energy splitting $\Delta E(\pi - \sigma)$ between the two components determines the environment effect on the atomic 1s → np resonance in linear and planar compounds.

Figure 11 illustrates the quasiatomic origin of the π and σ bands in 1s-shell absorption spectrum of the model N ---N system. The N 1s → 2pπ and 1s → 2pσ transitions are marked as arrows in the correlation diagram that links the limiting cases of the diverse N atoms (when $R \gg R_e$ and $\lambda \ll 1$, right-hand site) and the united atom Si (when $R \ll R_e$ and $\lambda \gg 1$, left-hand site). The splitting energy $\Delta_{\pi\sigma}^{2p}$ is also shown in the figure. The right panel displays the spectral distribution of oscillator strength corresponding to the $1s^{-1}\pi^*$ and $1s^{-1}\sigma^*$ resonances. The energy separation of the $1\sigma_g$ (1s'+1s'') and $1\sigma_u$ (1s'-1s'') molecular orbitals in ground state of the model system is not shown in the scheme.

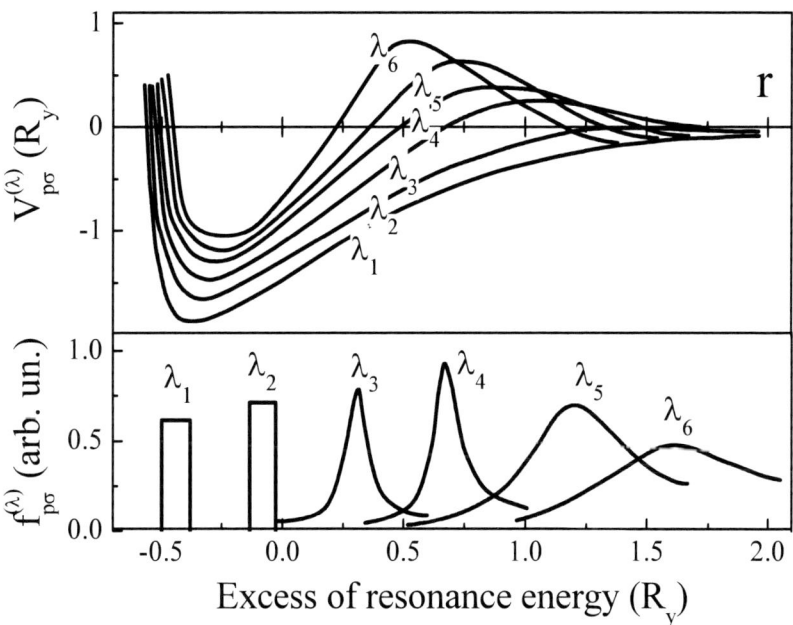

Figure 10. The dependence of the effective potential $V^{(\lambda)}(r)$ (upper plot) and the oscillator strength spectral density (bottom) on the parameter λ for boron atom in the basal plane [55]. $0 < \lambda_1 < \lambda_2 < \lambda_3 < \lambda_4 < \lambda_5 = 1 < \lambda_6$.

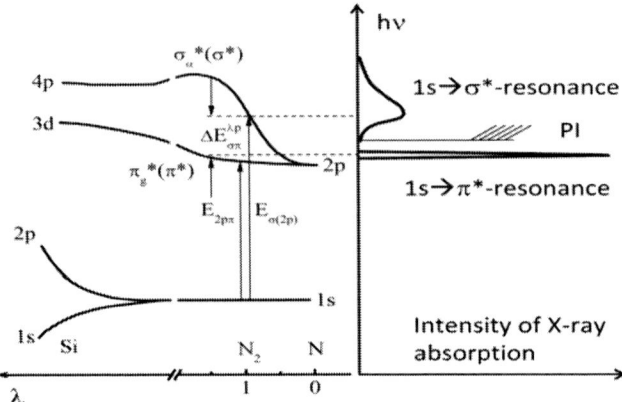

Figure 11. The splitting of the atomic 1s → 2p resonance in X-ray absorption spectra of linear (or planar) compounds. The left panel presents the splitting in the N_2 molecule ($\lambda = 1$) in ground state. The separated N atoms appear on the right-hand site ($\lambda = 0$) of the scheme and the united atom (Si) appears on the left site ($\lambda = \infty$) of the scheme. The right panel shows the spectral distribution of oscillator strength corresponding to the splitting. The 1s → π* and 1s → σ* resonances dominating in K-shell absorption of N_2 are shown.

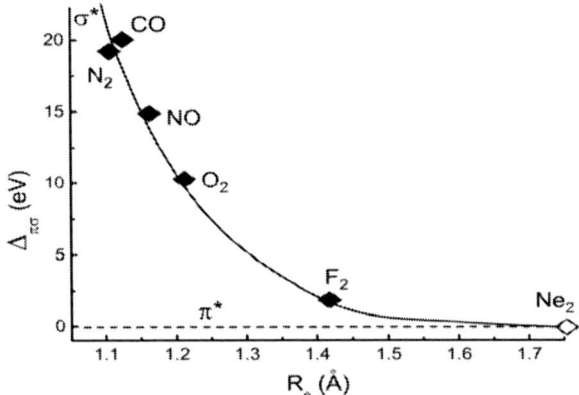

Figure 12. Dependence of the energy splitting $\Delta_{\pi\sigma}^{2p}$ on equilibrium distance in the row of diatomic molecules: N_2, CO, NO, O_2 and F_2. The positions of the σ* shape resonance relative to the lowest π* resonance extracted from the K-shell absorption molecular spectra are shown. For F_2 the separation is obtained by considering the energy separation between the high occupied π*-state and the σ*-resonance. As a limiting case $\Delta_{\pi\sigma}^{2p} \approx 0$ at $R_e \to \infty$ is added (Ne_2).

The QA approach to X-ray processes in polyatomic compounds [54, 55] allows one to describe X-ray absorption in terms of strongly localized atom-like functions [63, 64]. To find the changes in spectral distribution of oscillator strength for X-ray transition induced by the environment of the core-excited atom the variable phase approach to potential scattering [65, 66] is applied. The spectral changes for a $n_0 l_0 \rightarrow$ El transition in "atom-in-compound" relative to the free atom are characterized by the modulating $M_{l\Gamma}(E)$ function, where Γ is an irreducible representation of the point-group symmetry of an excited atom in a compound and E is the kinetic energy of the ejected photoelectron. Spectral dependence of the modulating function reproduces the main details of absorption spectra originated from the interference of the primary photoelectron ejected from the atomic core with the electronic waves scattered on the surrounding atoms.

To have an empirical evidence of the 1s $\rightarrow$ 2p resonance splitting $\Delta_{\pi\sigma}^{2p}$ we present dependence of the energy separation of the π and σ molecular states in the row of the diatomic molecules N_2, CO, NO, O_2 and F_2 on their equilibrium distances. Figure 12 displays the dependence. One may see that the energy $\Delta_{\pi\sigma}^{2p}$ approaches to nil with an increase in R_e. In the figure Ne_2 provides a limiting case of large interatomic distances.

In order to explain the extended dependence $\Delta_{\pi\sigma}^{2p}(R_e)$, which includes the different molecules, we attract attention to the following key points. For short R_e the σ^* resonances can be described as localized within the potential boxes and the energy of the resonances is determined rather by the size of the box than by the "contents". These points indicate on a correlation between the observed dependence and the results obtained by Sheehy et al. [122] for diatomic and pseudo-diatomic molecules in the framework of the "particle in cylindrical box" model. These points agree with the FEMO approach [123, 124].

The modulating function is determined as:

$$M_{l\Gamma}(E) = \text{Re}\left(\frac{1+BS}{1-BS}\right)_{l\Gamma,l\Gamma} \qquad (2)$$

B and **S** are the photoelectron reflection and scattering matrixes, respectively. $\mathbf{B} = \{B_{ll'\Gamma}\}$ is the main local electro-optical characteristic. It specifies the surrounding influence on the atom-absorber. Its elements are the amplitudes of back-scattered (reflected) photoelectron waves where l and l' are the orbital momentums of electron waves before and after their interaction with anisotropic and radial inhomogeneous surrounding potential. The matrix **B** obeys the non-linear first order differential equation (phase equation) [54, 65, 67]:

$$\mathbf{B}'(r) = [\mathbf{\Phi}^+ + \mathbf{B}(r)\mathbf{\Phi}^-]\mathbf{W}(r)[\mathbf{\Phi}^+ + \mathbf{\Phi}^-\mathbf{B}(r)], \qquad (3)$$

with the border condition $\mathbf{B}(E, R_F) = 0$, where R_F specifies the region inside which the reflected electron waves are formed. $\mathbf{\Phi}^\pm$ are diagonal matrixes, their non-zero elements being the partial radial wave functions outgoing (+) and incoming (-) into the excited atom, respectively. The asymptotes of the atomic functions are $\exp[\pm i(kr - l\pi/2 + C)]$ where C is the Coulombic phase. $\mathbf{B}'(r) = \{\frac{\partial B_{ll'\Gamma}}{\partial r}\}$ and **W** is the interaction matrix. Its elements are

$$W_{ll'\Gamma}(r) = \sum_{\mu \geq 0} W_\mu(r) \int Y^*_{l\Gamma}(\Omega) Y_\mu(\Omega) Y_{l'\Gamma}(\Omega) d\Omega, \qquad (4)$$

where W_μ is the 2^μ-pole momentum of surroundings potential $W(r)$. $Y_{l\Gamma}$ is a symmetry adapted linear combination of spherical functions centered at the excited atom. The surrounding potential $W(r)$ is defined as a superposition of atomic pseudopotential. Using eqs. (3) and (4) one may compute $M_{l\Gamma}(E)$ that is compared with experimental spectra. It is necessary to mark that the experimental partial X-ray absorption spectra are usually dominated by interference of the primary and scattered electron waves and multielectron excitations that are often intense near the thresholds. The modulating $M_{l\Gamma}(E)$ function describes specifically the interferential effect on photoabsorption.

The amplitudes $T_{ll'\Gamma}$ of electronic waves transmitted through the surroundings obey the second phase equation [54, 65, 67].

$$\mathbf{T}' = -\mathbf{\Phi}^+ \mathbf{W}[\mathbf{\Phi}^+ + \mathbf{\Phi}^-\mathbf{B}]\mathbf{T} \qquad (5)$$

with the border condition $\mathbf{T}(E, R_{core}) = \delta_{ll'}$. It shows that had the potential of the surroundings have been ignored, the outgoing photoelectron waves would have appeared in two (or one) dipole-allowed channels with $l = l_0 \pm 1$ only. This means that all other harmonics with $l' \neq l$ in the reflection matrix $\mathbf{T} = \{T_{ll'\Gamma}\}$ appear as a result of El-electron transmission through the anisotropic surroundings potential and $\mathbf{T}'(r) = \{\dfrac{\partial T_{ll'\Gamma}}{\partial r}\}$. The matrices $\mathbf{B}(E, R_{core})$ and $\mathbf{T}(E, R_F)$ determine completely the photoelectron wave functions. The matrices are computed by solving the nonlinear first-order (Riccaty-type) differential equations [54]. The electron reflectivity $\mathbf{R}(E)$ and transmission $\mathbf{T}(E)$ of the surroundings are defined as $\mathbf{R}(E) = |\mathbf{B}(E, R_{core})|^2$ and $\mathbf{T}(E) = |\mathbf{T}(E, R_F)|^2$. Evidently, without inelastic losses

$$\mathbf{R}(E) + \mathbf{T}(E) = 1 \qquad (6)$$

At large distances beyond the sphere of the radius R_F, the amplitudes of the reflected waves are negligible and the photoelectron wave function $\psi(E, r)$ can be presented as

$$\psi(E,r) = \sum_{l'\Gamma}[T + BST + ... + (BS)^n T]_{ll'\Gamma}\, \varphi_{l'}^+(r) Y_{l'\Gamma}(\Omega) \qquad (7)$$

$$\sim \sum_{l'\Gamma}\left[\dfrac{T}{1 - BS}\right]_{ll'\Gamma} \dfrac{1}{r} \exp[i(kr - \dfrac{l'\pi}{2})] Y_{l'\Gamma}(\Omega)$$

Using the expression for $\psi(E, r)$ a full dynamical (i.e. multiple-scattering) description of the photoelectron flux $J = \psi \nabla \psi^* - \psi^* \nabla \psi$ emitted from atomic core and having gone through the surroundings (i.e. for $r > R_F$) is provided.

Considering the photoelectron fluxes $J(E)$ crossing the spheres with the small radius R_{core} and the large radius R_F (see, figure 12) we define respectively the single-hole-creation (SHC) $\sigma^\oplus(E)$ and single-hole-ionization (SHI) $\sigma^+(E)$ cross sections. The cross section $\sigma^\oplus$ describes X-ray absorption and ejection of the photoelectron with kinetic energy E from the atomic core and σ^+ describes X-ray absorption and detection of the photoelectron far from the core ionized atom. In the framework of the fixed-nuclei approximation they can be computed as

$$\sigma^{\oplus}(E) = \sigma_0(E)\operatorname{Re}\left(\frac{1+BS}{1-BS}\right)_{l\Gamma,l\Gamma} \qquad (8)$$

and

$$\sigma^{\div}(E) = \sigma_0(E)\sum_{l'\Gamma}\left[\frac{T}{1-BS}\right]^2_{ll'\Gamma} \qquad (9)$$

In case $\nabla J = 0$ the condition (6) is valid and the SHC and SHI cross sections coincide one to another

$$\sigma^{\oplus}(E) = \sigma^{\div}(E) \qquad (10)$$

Below as a rule the SHC cross section is used. By comparing eqs.(8) and (9) we see that, in the first, $\sigma^{\oplus}$ is determined exclusively by the reflection matrix. This means that eq. (2) describes completely the surroundings effect on SHC. Unlike $\sigma^{\div}$ is determined by the both reflection and transmission matrices and solving of both eqs. (2) and (4) are required. In the second, the fluxes $J(E, R_{core})$ and $J(E, R_F)$ demonstrate different partial wave's compositions. According to eq. (8) only the dipole-allowed harmonics control the flux through the atomic core. In contrast a large number of l'-harmonics determine the flux crossing the sphere with the large radius. The number l_{max} depends substantially on E and R_F.

To help in understanding of the electron-optical characteristics of the surroundings figure 13 displays the spheres with the small and large radii (R_{core} and R_F) around the core-ionized atom in h-BN (in a basal plane) and presents schematically the surroundings potential $W(r)$ with dotted line. The core-ionized atom is placed in the center. The amplitudes B determined at $r = R_{core}$ are the superimpositions of incoming waves backscattered on the assembly of the surrounding atoms located inside the short range order region $R_{core} < r < R_F(k)$. The distant atoms located at $r > R_F$ do not influence on B.

Thus they do not play any role in forming spectral dependence of the SHC cross section.

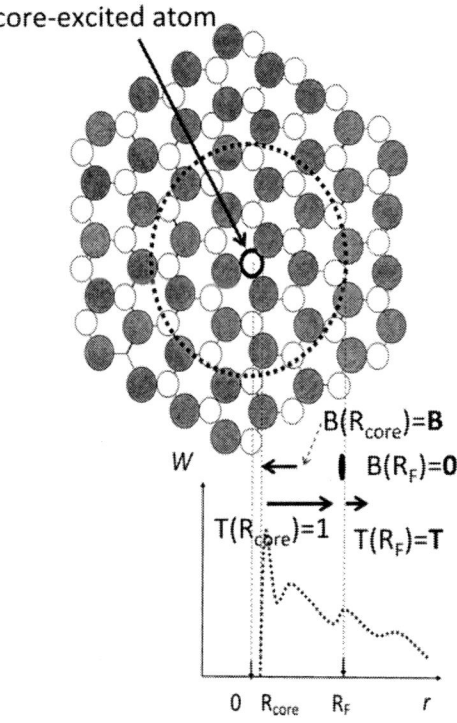

Figure 13. Formation of the reflection B and transmission amplitudes of the photoelectrons ejected from core-excited atom. The spheres of the small and large radii in the basal plane of h-BN are shown. The surroundings potential (monopole term) W(r) is also shown with dotted line.

For deeper understanding of the main relationships between X-ray absorption or inner-shell photoemission processes and the local electro-optical characteristics of matter let us consider the solutions of eq. 3 and 5 [125]. Evidently that for fast photoelectrons ejected from atomic core the amplitudes of reflected (back-scattered) waves are weak $\left|B_{//\Gamma}\right| \ll 1$. Then, neglecting by the terms dependent on the amplitudes in the right-hand sites of the equations we come to their kinematic approximation (*ka*):

$$\mathbf{B'} = \mathbf{\Phi}^+ \mathbf{W} \mathbf{\Phi}^+, \tag{11}$$

and

$$\frac{T'}{T} = -W\Phi^+\Phi^-. \tag{12}$$

By solving them we have

$$B^{ka}{}_{ll'\Gamma} \approx i^{l+l'+1} \int_{\Omega_s} W_{ll'\Gamma}(r)\exp(2ikr)dr \Big/ 2k \tag{13.1}$$

and

$$T^{ka}{}_{ll'\Gamma} \approx \exp(i^{l-l'+1} \int_{\Omega_s} W_{ll'\Gamma}(r)dr \Big/ 2k. \tag{13.2}$$

The surrounding region is denoted via Ω_s. Within the kinematic approximation phase shifts characterize the electron transmission through the surroundings as $\left|T^{ka}{}_{ll'\Gamma}\right| = 1$. The radii of the surrounding regions for the amplitudes $B^{ka}{}_{ll'\Gamma}$ and $T^{ka}{}_{ll'\Gamma}$ differ. In particular, for fast photoelectrons ($k \gg 1$) due to quick oscillations of the factor e^{2ikr} in eq. 13.1 the surrounding region Ω_s assisting in formation of $B^{ka}{}_{ll'\Gamma}$ is much narrower than the region assisting in formation of $T^{ka}{}_{ll'\Gamma}$.

When the amplitudes $B_{ll'\Gamma}$ are weak but not negligible we keep in the right-hand site of eqs. (2) the terms dependent linearly on **B** too. Then, we have

$$\mathbf{B'} = \mathbf{\Phi^+ W\Phi^+} + \mathbf{B\Phi^+\Phi^-} + \mathbf{\Phi^+ W\Phi^- B}. \tag{14}$$

Neglecting by non-diagonal elements $B^{ka}{}_{ll'\Gamma}$ and using the asymptotes of the functions $\varphi^{\pm}_{l'\Gamma}$ one may present the solution of the linearized eq. (14):

$$B^{ka}{}_{ll'\Gamma} \approx i^{2l+1}/2k \int_{\Omega_s} W_{ll'\Gamma}(r)\exp(2\frac{k}{n_{l\Gamma}}r)dr \ . \tag{15}$$

The local index of electron refraction $n_{l\Gamma}(k)$ is

$$n_{l\Gamma}(k) \approx 1 + \frac{<W_{ll'\Gamma}(r)>}{2k^2} \ . \tag{16}$$

$<W_{ll'\Gamma}(r)>$ is the averaged surroundings potential proportional to $1/r \int_{R_{core}}^{r} W_{ll'\Gamma}(r')dr'$. When the amplitudes $B_{ll'\Gamma}$ are high the nonlinear term in the right-hand site of the eq. (2) cannot be ignored. Regrettably analytical solutions of eq. (3) do not exist excepting the case when the δ-potential approximation to the surroundings potential is used. Presenting the surroundings potential $W(r)$ as a superposition of Dirac's bubble potentials $\sum_j w_j \delta(r - R_j)$ where j enumerates the coordination atomic shells with the radii R_j and scattering capability w_j, one may present the solution of eq. (3) by using the recurrent relation [54, 60, 72, 73, 126]

$$B_\Gamma \approx \frac{b_j + q_j b_{j+1}}{1 - a_{jj+1}} \ , \tag{17}$$

b_j is the amplitude $|b_{jl\Gamma}|\exp(2ikR_j + \phi_{jl\Gamma})$ describing the reflection from j-th coordination atomic shell taken as isolated one. From equation (17) follows that $B_{l\Gamma}$ cannot be presented by a superposition of independent waves backscattered from the separated j-th shells. Interference of the scattered waves results in the interdependence, where $a_{jj+1} = b_j b_{j+1} \exp(-4ikR_j)$ and $q_j = -\exp(4i\phi_j)$. The interdependence becomes negligible when both b_j and b_{j+1} are small. We note here that the amplitude $B_{l\Gamma}$ found as a solution of eq. (3) coincides with the amplitude found by considering the multiwaves interference of electronic waves within j-th and $j+1$-th coordination shells. This allows us to consider the surroundings similar to a nanoscaled Fabri – Perreault interferometer. Spectral dependences of $B_{l\Gamma}$ and $M_{l\Gamma}$ that take the interferential effects into account are discussed in more detail in [54, 60, 72, 73, 126].

DYNAMIC LOCALIZATION OF CORE-EXCITED STATES

X-ray absorption and inner-shell photoionization encounter great difficulties with the description of core-excited states in compounds with equivalent atoms. Are B and N 1s holes spatially localized or delocalized? Core-hole localization means that the translation and inversion symmetries are broken in the ground state. Hence the core-excited atom can be regarded as a defect center in the lattice. In contrast, if core-hole is regarded as delocalized symmetries of ground and core-excited states coincide. This is a long-living problem in x-ray spectroscopy.

The equivalence of atomic sites in a polyatomic system implies their equal probability of excitation (the value averaged over large timescale) but not simultaneous core excitation. This means that one-photon absorption of the quasidegenerate 1s levels occurs in one of the equivalent atoms in a system, and the photoelectron wave function Ψ should be presented as a symmetry-adapted linear combination of atomic functions φ_n, which describe photoelectrons emitted from n-th equivalent position in a system. To take this dynamic localization into account, different time dependence for the atomic wave functions is assumed in this superposition:

$$\Psi(t) = \sum_n c_n \varphi_n (t - t_n - \tau). \tag{18}$$

For the first time this time dependent approach was successfully applied to describe the photoelectron angular distributions from N and O 1s levels in fixed-in-space N_2 and CO_2 molecules [127]. Evidently that $|c_n| = 1$ and t_n is the beginning of core-ionization. Assuming that t_n and $t_{n\pm 1}$ are essentially different $|t_n - t_{n\pm 1}| \gg T$, where T is a time characterizing the interaction of photoelectrons with the anisotropic molecular (or cluster) potential, we ignore the interference of the φ_n and $\varphi_{n\pm 1}$ waves. This means that K-shell photoemission is composed from the independent photoelectron fluxes outgoing from all equivalent atoms and the total flux is

$$J \propto \sum_n |\varphi_n|^2 . \qquad (19)$$

One can assign the φ_n functions to incoherent waves. This particular formulation of K-shell photoemission makes possible to consider a core-excited atom as a point defect center in the crystalline lattice. The inversion symmetry is broken.

A quasistationary description supposes that equivalent atoms must be treated as equivalent sources of electron waves, and the relevant wave function $\Psi(t)$ is given by eq.(19) at $t_1 = t_2 = \ldots = t_n$. The emission occurs simultaneously in all equivalent atoms in a system with equal probability, and the process is described as being quasistationary. Since all equivalent sources are located in the same lattice, the φ_n waves are coherent and give the flux

$$J \propto |\sum_n c_n \varphi_n|^2 . \qquad (20)$$

The interference term $2\sum_{n,n' \neq n} \mathrm{Re}\,\varphi_n^* \varphi_{n'}$ makes the difference between the localized and delocalized descriptions.

A question now arises: are equivalent atoms coherent or incoherent sources of electron waves in a system? We stress that the same question arises in molecular studies, e.g. examining N and O 1s photoemission from N_2 or CO_2. Many efforts are invested to solve it (see, e.g. [127 – 130]). There are evidences indicating both strong spatial localization and delocalization of inner-shell processes in matter. Investigations of free atomic and molecular clusters make possible to link the core-hole localization problem in free molecules and solids.

To answer this question another time (τ) dependence associated with core-hole relaxation is introduced. It takes changes in core-hole localization into account. τ is determined by the hopping time t_h and the timescales t_e and t_r, which are characterized by the core-hole electronic decay and the relaxation of the nuclear subsystem, respectively. In addition, we introduce the trapping time T of the photoelectron within the short range molecular (or cluster) potential. Approximately

$$T \propto \frac{\mu R_{trap}}{v},$$

μ is the multiplicity of photoelectron scattering on neighboring atoms, v is the photoelectron velocity and R_{trap} is the radius of trapping region. Thus a competition between the various timescales ($t_n - t_{n'}$, t_h, t_e, t_r and T) controls the dynamic properties of the photoemission processes. Figure 14 displays schematically the core-hole dynamic localization in a cyclic system in supposition that $|t_n - t_{n'}| \gg \max\{t_h, t_e, t_r, T\}$. The figure shows that at $t = t_1$ (or t_2) a single x-ray photon is absorbed by the system possessing ground state symmetry. The following step: the ground state symmetry is reduced due to ejection of core-electron and creation of core-hole in one of equivalent atoms in the system. The following step: core-hole decay and core-hole hopping processes restore the ground state symmetry.

Two limiting cases can be distinguished:

$$T \ll \tau$$

and

$$T \ll \tau.$$

The former inequality is predominantly valid for core-excitation continua, where fast photoelectron is emitted. It describes strong core-hole localization in one of equivalent atoms. Therefore the electron flux is described by eq.(19). In contrast, the latter inequality describes fast core-hole delocalization, e.g. due to fast core-hole hopping or $Core^{-1}VV$ Auger decay. Hence the symmetry of the excited system is quickly restored. Equivalent atoms are regarded in this case as coherent sources of photoelectron waves. Hence no contradiction between the localized and delocalized descriptions appears if core-excitations are regarded as time-dependent processes. In general, the electrons move in a time-dependent potential of varying symmetry which cannot be assigned to a fixed point group.

We also stress that core-hole electronic relaxation tends mainly to restore the ground state symmetry. In contrast, relaxation of nuclear subsystem tends to retain the broken local symmetry. In the first, the creation of electron-hole pair disturbs the balance of electrostatic forces keeping the ground state symmetry. The residual asymmetric force arising simultaneously with the core-hole, leads to asymmetric atomic motions. In the second, the ejection of an electron from atomic core is accompanied by the local recoil resulting in asymmetric motion too.

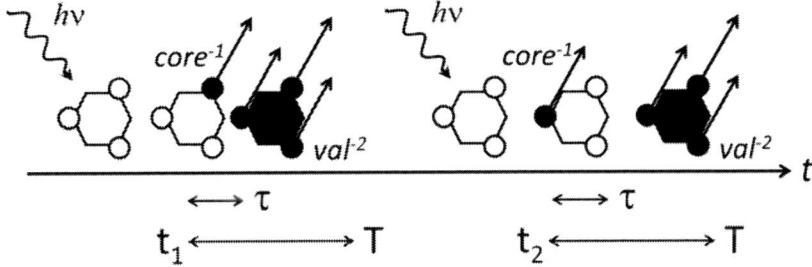

Figure.14. Correlation of core-electron emission from a core-excited molecule with core-hole relaxation. Outgoing photoelectrons are shown are shown by arrows with inclusion of multiple scattering.

By using photoelectron – photoion coincidence technique K-shell excitation dynamics is studied in more detail in small molecular species than in solids.

PHOTOELECTRON LOSSES AND SHAPE RESONANCE PHENOMENA

Up to now the photoelectron losses in the surroundings are neglected. This is a crude approximation. In addition to the elastic photoelectron scattering on neighboring atoms there are inelastic scattering effects. In the framework of the quasiatomic approach these effects can be taken into account by using the concept of the optical potential [131 – 132]. The surroundings optical potential can be introduced as [54, 133]

$$W^{opt}(r) = V(r) - iU(r). \qquad (21)$$

Positive U values refer to dissipation of the photoelectron flux in the surroundings (see, figure 12). As a result, the equality (6) becomes invalid. From equations (9) and (10) it follows that $\sigma^{\oplus} > \sigma^{+}$ and

$$\sigma^+(k) \approx e^{-x}\Lambda(k)\sigma^\oplus(k). \qquad(22)$$

Here $\Lambda = (1+z^2 Z)[Z(1+z^2)]^{-1}$, where $x = U/k$, $z = k/W$, $y = xz$ and $Z = (1+x)(1+y^2)^{-1}$. The analysis shows that apart the low k regime the exponential damping of the photoelectron flux dominates the ratio $\sigma^+/\sigma^\oplus$. But for low k the Λ function becomes important, sharply reducing the ratio as $\Lambda \to 0$ at $k \to 0$.

In addition to the deviation between the σ^+ and $\sigma^\oplus$ cross sections the imaginary part results in appearance of the additional phase shift $\beta < 0$ of the photoelectron wave [133]. Due to the phase shift the moving of the photoelectron has to be regarded as correlated by valence excitations. The photoelectron – valence electrons correlations change the spectral dependence of the single-hole creation $\sigma^\oplus$ and single-hole ionization σ^+ cross sections. Within the optical potential concept and the quasiatomic approach the photoelectron – valence electrons correlations lead to (a) the decrease of σ^+ right above the inelastic threshold, (b) an anomalous spectral behavior of the both cross sections at the threshold and (c) an upward shift of the shape resonance [133]. The experimental studies performed in free molecules [77, 134, 135] and molecular clusters [72, 73] support these general conclusions.

Shape resonances are well known in core-level excitation spectroscopy [54 -56, 71 – 74, 136 – 138]. Often they are treated as one-electron phenomenon. The resonances dominate the x-ray absorption and inner-shell photoemission of free molecules, clusters and solids. A shape resonance in inner-shell photoemission is associated either with temporary trapping of an electron ejected from a deeply bound level by the finite-size potential barrier and subsequent tunneling of the photoelectron through the barrier into the core ionization continuum or with transition from a core levels to virtual molecular orbital lying in core ionization continuum [170-172]. The shape resonance process is usually described in terms of intramolecular interferences of scattered photoelectron waves. Kronig [169] suggested long ago that the outgoing electron ejected from atomic core interferes with the waves scattered by other atoms in a compound (originally in a molecule). This interference produces a $\sin(2kR)$-modulation factor, where k is the photoelectron wave number. For low k the multiple scattering of the ejected electron increases and results in appearance of shape resonances. The resonances in figure 10 corresponding to the parameters $\lambda_{n\geq 3}$ can be assigned with the shape resonance features in spectral distribution of oscillator strength.

For many years their substantial R-dependence has attracted much attention as a source of structural information [136 – 138] (see, also below). However, the accurate extraction of bond lengths encounters great difficulties, mainly because of valence excitations yielding an increased uncertainty in the determination of the energy position of the shape resonances. In addition to this, the ejected electron couples with valence electrons resulting in the distortions of both line shape and position of a shape resonance [77]. Vibrational excitations associated with inner-shell ionization also influence the shape resonance position [78].

Using eq.(8) we represent the interferential function as

$$M_{I\Gamma}(k) = \text{Re}\left(\frac{1+BS}{1-BS}\right)_{I\Gamma,I\Gamma} = \frac{1-\rho_{I\Gamma}^2(k)}{1+\rho_{I\Gamma}^2(k)-2\rho_{I\Gamma}(k)\cos 2\eta_{I\Gamma}(k)} \quad (23)$$

The phase shift $2\eta_{I\Gamma}(k)$ is equal to $2kR + 2\delta_l(k) + \varphi_{I\Gamma}(k)$, where δ_l is the phase shift of electron scattering at the core-excited atom, $\rho_{I\Gamma}$ and $\varphi_{I\Gamma}$ denote the reflection coefficient and the phase shift of the total back scattered wave, respectively. $\rho_{I\Gamma}$ and $\varphi_{I\Gamma}$ are expressed as combinations of $|B_{II'\Gamma}|$ and $\arg(B_{II'\Gamma})$. The function $M_{I\Gamma}$ can be also expanded in the Tchebychev polynomial series [139]:

$$M_{I\Gamma} = \sum_{m\geq 0} T_m(z) = 1 + 2\sum_{m\geq 1} \rho_{I\Gamma}^m \cos(2m\eta_{I\Gamma}). \quad (24)$$

Here $T_m(z) = \cos(m \arccos(z)) = d^m(1-z^2)^{m-\frac{1}{2}}/dz^m$. The condition $\partial M_{I\Gamma}(k)/\partial k = 0$ leads to the phase shift equation:

$$2kR + 2\delta_l(k) + \varphi_{I\Gamma}(k) + \Delta(k) = \pi N \quad (25)$$

Even N=2n gives k_n referring to maximums of $M_{l\Gamma}$. The phase shift $\Delta(k)$ is tightly related with spectral variations of $\rho_{l\Gamma}$. Usually $\left|\partial\rho_{m\Gamma}/\partial k\right| << \left|\partial\eta_{m\Gamma}/\partial k\right|$ hence eq.(25) can be approximated

$$2kR + 2\delta_l(k) + \varphi_{l\Gamma}(k) \approx \pi N \qquad (26)$$

For the shape resonance features N = 0 or 2. Eq.(26) coincides with that of derived in the works [140, 141].

When photoelectron losses are taken into account the additional phase shift β appears. $\beta < 0$ for $E > E_{inel}$ where E_{inel} is the threshold of photoelectron inelastic losses. Then, eq.(26) can be rewritten

$$2kR + 2\delta_l(k) + \varphi_{l\Gamma}(k) + \beta(k) \approx \pi N \qquad (27)$$

As an illustration figure 15 shows the characteristic changes in the shape resonance vicinity for both σ^+ and $\sigma^\oplus$. One may see the deviation of the single-hole creation and ionization cross sections above the inelastic threshold energy (E_{inel}) and the irregular behavior i of the both cross sections at the energy.

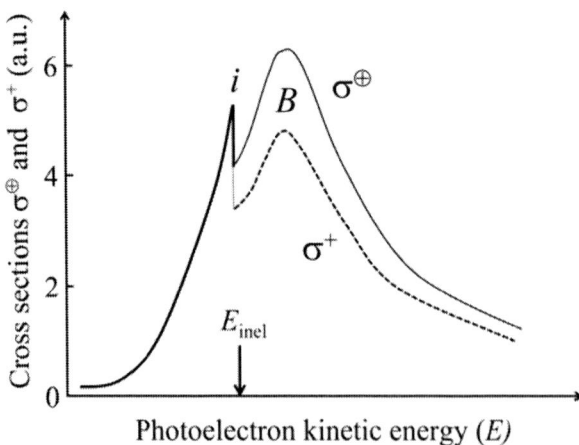

Figure 15. Characteristic line shape of a shape resonance in respect to the photoelectron – valence electrons coupling. Inelastic threshold for valence excitations is shown by arrow. $\sigma^+ = \sigma^\oplus$ for $E < E_{inel}$ (as $U = 0$) and $\sigma^+ < \sigma^\oplus$ for $E > E_{inel}$ (as $U > 0$).

When $U\equiv 0$ the shape resonance is described as a result of interference of the photoelectron waves, occurring in the rigid compound (valence electrons are independent spectators of the photoelectron motion). The interferential picture changes when $U > 0$. The importance of the photoelectron – valence electrons coupling suggests that one describe the interference as occurring in the soft compound. This means that due to the coupling the valence excited configurations assist in the interference. The molecular study [77] evidences that the coupling plays a significant role in forming the shape resonance profile.

Chapter 4

BORON AND NITROGENE EXCITATIONS

BORON EXCITATION

Let us consider B K shell excitations in h-BN. The experimental B K-shell absorption spectrum of h-BN crystal measured with very high energy resolution [59] is plotted in figure 16 (*a*). The absorption spectrum demonstrates a very narrow (≈ 0.37 eV) and intense discrete peak *A* at 191.8 eV and an intense band *B* with two maxima at 197.9 eV and 199.2 eV. The absorption cross section oscillates for higher photon energies coming through maxima at 204.1, 207.4 and 215.6 eV.

The calculated modulation functions $M_{a_2^"}$ and $M_{e'}$ describe the spectral distributions of the dipole allowed $B1s \rightarrow Ep_z(a_2^")$ and $B1s \rightarrow Ep_{x,y}(e')$ transitions. Axis *z* is oriented perpendicular to the basal plane (*x*, *y*). The energy-dependent model Heine-Abarenkov potential [68] for the nearest neighbors in the plane and zero-range potential approximation [69,70] for distant atoms in the crystal is applied to construct the surroundings potential $W(r)$. The hybridization of *p*- with higher *l*-partial waves is ignored. In the framework of these approximations $M_{a_2^"}$ is close to unity. This means that the spectral distribution of oscillator strength for $B1s \rightarrow Ep_z(a_2^")$ transition approaches to the atomic-like distribution. On this background the lowest intense peak A is assigned approximately with the boron $1s \rightarrow Ep_z(a_2^")$ excitation.

In contrast, the surroundings potential in the basal plane is essential and deforms substantially the spectral distribution of oscillator strength for doubly degenerated $1s \rightarrow Ep_{x,y}(e')$ transitions in a boron atom. To describe the surroundings effect in the basal plane the cluster approximation $B^+N_3B_6$ is applied. However, according to the calculations the reasonable description of the B 1s excitations can be obtained by using the small $[B^+N_3]$-cluster approximation.

Figure 17 exhibits the calculated spectral distribution of oscillator strength for transitions from B 1s level [58]. Comparison of the experimental and theoretical spectra makes possible assigning the features A and B in figure 16 with the resonances in single $B1s \rightarrow Ep_z(a_2'')$ and double degenerated $B1s \rightarrow Ee'(2p_{x,y})$ transitions, respectively. Thus, the spectral features A and B can be attributed to the splitting components of the B 1s → 2p resonance. Below we discuss the splitting value (i.e. the energy separation between the "π" and "σ" resonances) $\Delta_{\pi\sigma} \approx 7$ eV, which is typical for this type of inner-shell excitations in planar compounds. $\Delta_{\pi\sigma}$ is the most important parameter characterizing the short-range order in h-BN.

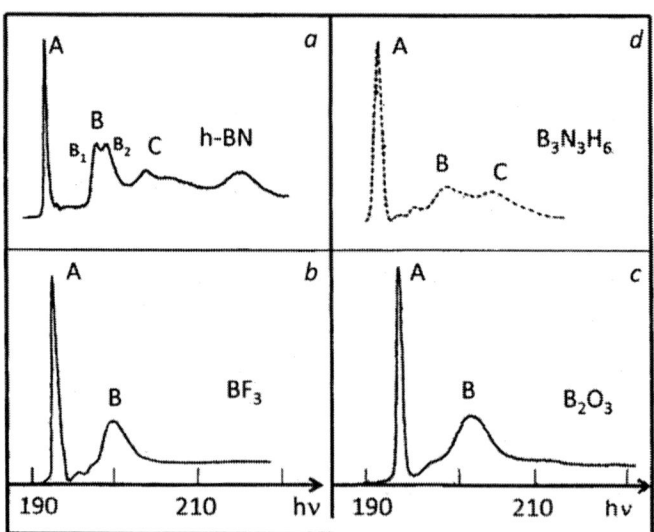

Figure 16. The B K-shell absorption spectra of h-BN (*a*) and B_2O_3 (*c*)[59] measured in the total electron-yield mode and BF_3 (*b*) [81] and $B_3N_3H_6$ (*d*) [142].

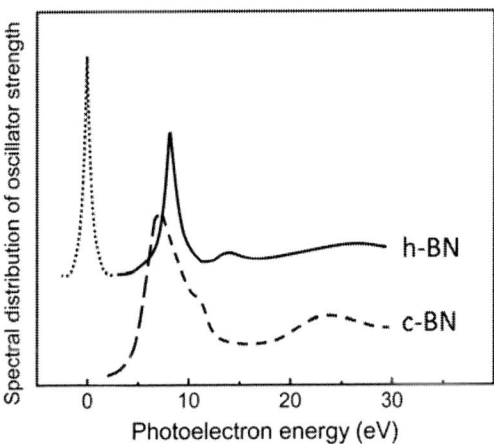

Figure 17. Calculated spectral distributions of oscillator strength from B 1s level into doubly degenerated σ-states (solid line) and π–state (dotted line). For comparison the spectral distribution for B 1s → t_2 transitions in cubic c-BN is also shown [58].

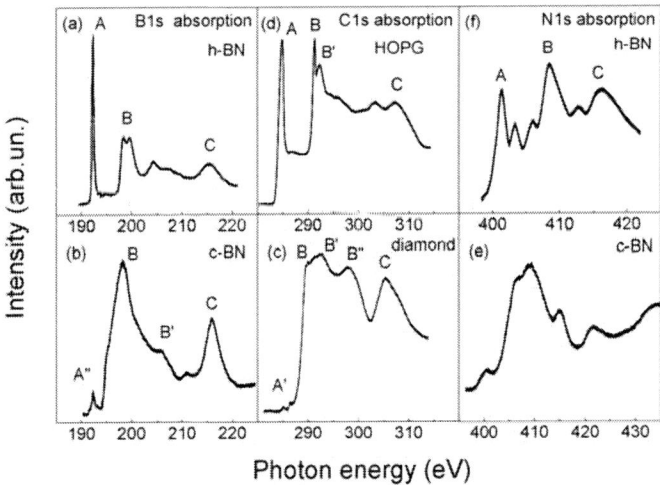

Figure 18. Experimental B, C and N K shell absorption spectra of c-BN, h-BN [58], graphite (HOPG) and diamond [143].

In the framework of the $[B^+N_3]$-cluster approximation to B K excitations the resonance B can be assigned with shape resonance phenomena. In more detail they are investigated for small molecular species (see, e.g. [71]). In particular, the close resemblance with the $B1s \rightarrow Ee'(2p_{x,y})$ resonance (i.e. the σ-component) in the free BF_3 molecule has to be mentioned. Usually (see chapter 3), shape resonance processes are associated with the trap of the photoelectron ejected from atomic core by the cluster (or molecular) potential barrier with the subsequent tunneling of the photoelectron through the barrier to the core-hole ionization continuum. In solids the essential difference with the molecular shape resonance process occurs in the continuum due to the photoelectron scattering at the distant atoms. In what extend the scattering influences on the photoelectron trap within the barrier? This problem is examined rather in detail in the works [72 -74].

The spectral features at 205 eV and 215 eV in figure 16 can be assigned with electron scattering on neighboring atoms in the basal plane. For comparison the computed spectral distribution of B 1s transitions in cubic-BN is plotted in figure 18, too. Specifically, there is no the π − σ splitting and the triply degenerated $B1s \rightarrow t_2(2p_{x,y,z})$ resonance dominates in the distribution in full agreement with the experimental spectra [58].

Figure 16 provides an experimental illustration of the splitting mechanism. To confirm the link of the spectral changes with the anisotropy within and perpendicular to the basal plane and with the nearest neighbors in the plane the B K shell absorption spectra of amorphous and planar B_2O_3 crystal [58] and the nearly planar $B_3N_3H_6$ and BF_3 molecules are plotted. Two resonance features A and B dominate in the spectra. Figure 18 provides an additional experimental support of the splitting mechanism. The B, C and N K shell absorption spectra of planar h-BN [58] and graphite (HOPG) [143], and cubic c-BN [58] and diamond [143] make evident the degeneration of the splitting components. In particular, the spectrum of c-BN is characterized by an intense absorption band at 197.4 eV referring to triply degenerated $B1s \rightarrow t_2(2p_{x,y,z})$ transitions. Close similarity with the C 1s absorption in diamond [143] can be mentioned. The peak A disappears. In general there is a well agreement with the measurements of Chaiken with co-workers [75, 76]. The close similarity in spectral distribution of oscillator strength for transitions B 1s shell in h-BN and amorphous B_2O_3 and the molecular species supports the applicability of the [BN_3]- and [BO_3]-cluster approximations to the description of the main spectral distribution of oscillator strength for B 1s transitions.

THE $\Delta_{\pi\sigma}$ SPLITTING

Since the π (A) -and σ (B) -resonances have a common atomic origin and they are expected to be localized predominantly in the planar $[B^+N_3]$-cluster one may consider their splitting energy $\Delta_{\pi\sigma}$ as a characteristic of local anisotropy of the electronic properties of the compound. Evidently, that the splitting energy $\Delta_{\pi\sigma}$ will depend on the anisotropy. In the works [58, 59] the splitting $\Delta_{\pi\sigma}$ is regarded as a measure of interatomic separation R between the core-excited boron and the nearest nitrogen atoms in h-BN. This treatment is discussed in more detail in the works.

Figure 19 taken from [59] presents the experimental dependence of $\Delta_{\pi\sigma}$ on the equilibrium interatomic B – X separation R_e. The dependence is extracted from B 1s spectra of h-BN and planar molecules BF_3, BCl_3, BBr_3, $B_3N_3H_6$, and planar BO_3^{3-} complex anions in the cubic crystals, where X = F, Cl, Br, N and O. The smooth and nearly linear experimental dependence of $\Delta_{\pi\sigma}$ (R_e) is seen unambiguously. The splitting in h-BN is marked by two symbols 3' and 3" referring respectively to the energy positions of the maximums B_1 and B_2 in figure 16 (a). Their origin is not clear now. Tentatively the double-top line shape of the resonance B can be assigned with the irregular behavior of the SHC cross section (see, figure 15) at the inelastic threshold lying close to the shape resonance energy.

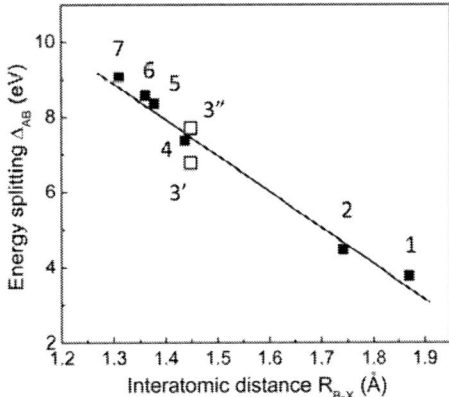

Figure 19. The dependence of the energy splitting $\Delta_{\pi\sigma}$ on interatomic B – X distance in various molecular species and solids extracted from the experimental spectra. (1) – (7) refer to BBr_3, BCl_3, h-BN, $B_3N_3H_6$, B_2O_3, borate BO_3^{3-}, BF_3, respectively. Symbols 3' and 3" mark the splitting energies referring to B_1 and B_2 in figure 16 (a).

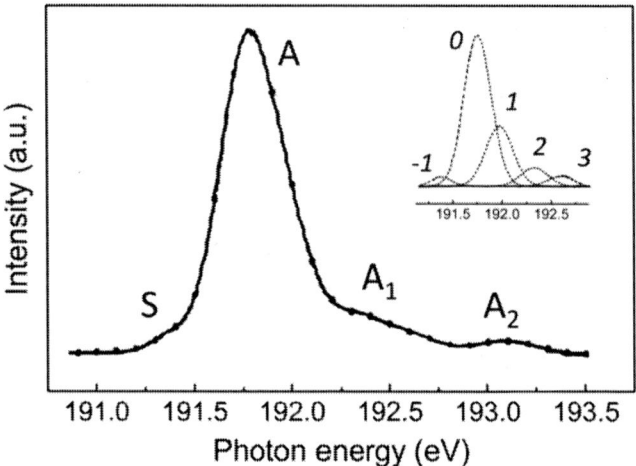

Figure 20. The $\pi^*(2pa_2'')$ resonance in B 1s shell absorption in h-BN [59]. Insert: the decomposition of the resonance. In addition to the main component marked as "*0*" extra components "*-1*", "*1*", "*2*", "*3*" appear.

The correlation of $\Delta_{\pi\sigma}$ with R plotted in figure 19 indicates both the strong spatial localization of the core-excitations within the molecular-like $[BX_3]^{n-}$ fragments and the parent relation of the π and σ excitations with the triply degenerated in a boron atom $1s^{-1}2p$ resonance. The linear dependence of the energy splitting on the interatomic distance marked with solid line provides evidently crude approximation to inner-shell excitations in the solids and for the molecular species too.

A question arises: in what extent the linear approximation works to use the experimental energy positions for the determination of the bond length? We note that this linearity is a subject of intense debate with both evidence favoring the so-called bond-length-with-a-ruler method and indications that its application to extract bond lengths from excess energy positions of the shape resonance is at best unsafe (see, e.g. [71]). The linear approximation $\Delta_{\pi\sigma}(R) \approx E_0(R_0) + \kappa\delta R$ reflects a linear response of the core-excitations on the environment. It is valid under the following conditions: (i) $\delta R/R = |R - R_0|/R_0 \ll 1$ and (ii) the splitting components (π and σ) are strongly localized inside the region of the core-excited atom and its nearest neighbors. In addition to the conditions there are at least three main sources leading to the violation of the linearity. They determine the possibility and the precision of the bond-length-with-a-ruler method. In the first, the resonance σ-component lying in continuum must be attributed to the shape resonance. This means that the multielectron excitations accompanying inner-shell ionization do not disturb it substantially (see, e.g. [71, 77]). In the second, the vibrational excitations accompanying inner-shell excitations must be taken into account. The investigations of the Franck-Condon effects on the shape resonances in molecular species evidence that spectral positions of the resonances cannot by unambiguously assigned to bond length because their positions depend on geometric changes in ground and core-ionized states [78, 79]. By analyzing inner-shell excitations in solids we encounter with similar difficulties because of strong core-hole localization. In the third, even if the photoelectron scattering at the nearest neighbors plays dominant role in forming the shape resonances and the scattering at the distant atoms is rather weak it cannot be completely ignored. Recently, it is shown [80] the scattering leads to spectral shifts of the molecular-like shape resonances in the molecular solids.

Let us consider the lowest absorption peak in more detail. The line shape analysis of the peak in figure 20 shows that its full width at half-maximum (FWHM) amounts to ~0.32 eV [59]. This is a typical value for the atomic X-ray transition in a solid and supports the conclusion [59] regarding its atomic origin and the non-bonding character of the relevant a"$_2$ MO. The FWHM is close to that of ($\approx$ 0.25 eV) observed in high resolution B 1s spectroscopic studies of the free BF_3 molecule. The line shape analysis performed in the work [59] shows that there is an additional structure consisting of a shoulder A' centered at 192.4 eV and a second peak A" at 193.1 eV.

The decomposition of the π-resonance shows a set of nearly equidistant discrete peaks labeled as 1, 2 and 3 lying at +0.26, 0.61 and 0.86 eV above the main transition centered at 191.75 eV. The experimental FWHM of the isolated peak at 193.1 eV coincides precisely with the FWHM of the main excitation. The recognized structure of the resonance is an argument against the conventional picture of phonon broadening of the X-ray transition in a solid.

The relationship between the energy splitting $\Delta_{\pi\sigma}$ on interatomic B – X distance in various molecular species and solids extracted from the experimental spectra plotted in figure 19 allows assigning the absorption band A and B in figures 16 and 18 to the splitting components of a parent atomic B $1s^{-1}2p$ resonance. By comparing the X-ray absorption near edge structure with the B K shell excitations in free planar BF_3, BCl_3 and BBr_3 molecules [81], the molecular origin not only of the A and B resonances but also of the outstanding peak A" in h-BN can be revealed. Ishiguro et.al. [81] have suggested that it is related to molecular dissociation. This suggestion agrees with the conclusion of Ueda [82] that when a B 1s electron is excited to the lowest orbital $2a_2^{''}$, the BF_3 molecule is expected to start deformation into pyramidal structure in C_{3v} symmetry (see, figure 21 (a)). Molecular dissociation measurements suggested this deformation [83, 84]. Theoretical calculations also suggested this possibility [83, 85].

Often quasi-molecular simulation is a useful tool for the description of electronic excitations in solids [54]. Due to strong spatial localization of electronic transitions such simulation can be successfully applied to atomic dynamics in clusters and solids [54, 72, 73, 74, 86]. In view of the strong spatial localization of the π-excitation in h-BN its molecular modeling is extended to atomic dynamics near the core-excited B atom in the crystal. Therefore, the spectral features *1, 2, 3* and *A"* in the B K absorption can be regarded as the result of local rearrangement in atomic structure near the excited atom [59]. This supposition correlates with the strong influence of vibronic coupling on the decay of the excitations [87] and with the essential anharmonicity of the vibrations [88] accompanying them. The harmonic approximation is certainly of limited applicability. Its correction has to include not only a cubic term in the lattice coordinates, but also asymmetric forces which can be especially important for excited quasi-two-dimensional systems. Then the relaxation (electronic decay and atomic rearrangement) of the X-ray excitation in h-BN can be approximately described as occurring within the BN_3-fragment. As a result it acquires molecular properties. Similarly to the case of molecular dissociation, the core-excited atoms leave their regular positions and make it possible to create defects in the lattice. This supposition correlates with the spectral features of X-ray emission from B_2O_3 associated with some small fraction of a lattice defect or impurities of boron atoms [89].

To describe the relaxation in B K excited h-BN, the methods developed to establish the core-level decay in molecular species (see, e.g. [90, 91]) can be applied. Then, the two timescales characterizing electronic and atomic relaxations near the core-excited atom can be introduced. There exist well known molecular examples where one or the other timescale is short enough to dominate the decay spectrum [90, 91]. However, for a lot of cases, a delicate balance between the two processes is required to describe the experimental data.

For the π-excitation in h-BN the electronic timescale is rather large and the atomic rearrangement in the vicinity of the core-excited atom cannot be ignored. With respect to the molecular (BF_3) dynamics near the $B1s^{-1}2p(a_2^{''})$ excitation [84, 92] characterized by the strong enhancement of the B^+ ion yield, the appearance of out-of-plane displacements of core-excited boron atoms seems to be rather reasonable. In particular, a transition from its planar (ground state, D_{3h}) configuration to pyramidal (C_{3v}) one is expected [59] (the both configurations are exhibited in figure 21 (*b* and *c*)). Such a pyramidal [B*N$_3$]-configuration (like planar-compressed one) can be regarded as a small-radius polaron accompanying the B K-shell excitation in the crystal. The out-of-plan displacement is estimated as ≈ 0.16 Å [59]. Nevertheless, we stress here that the extraction of structural information from either the σ-π splitting or shape resonance energy remains a complex problem that requires further theoretical and experimental investigations.

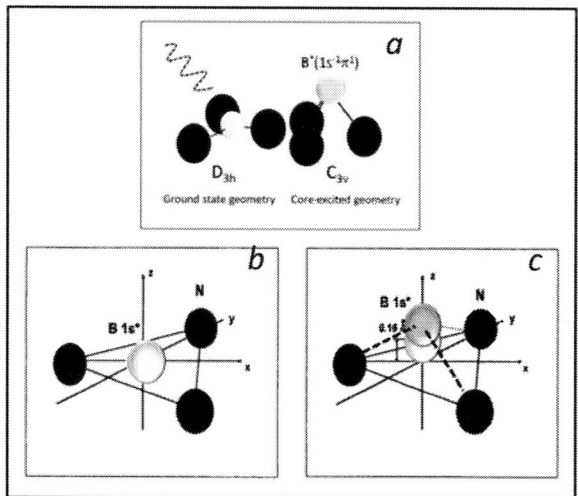

Figure 21. The pyramidal deformation in molecular BF3, associated with the B1s → $2a_2^{''}$ excitation (*a*). The planar (*b*) and the pyramidal (*c*) [BN$_3$]-configuration referring to the B K excited states at A and A″ bands. The out-of-plane displacement ≈ 0.16 Å of the boron atom is determined in the work [59] by using the $\Delta_{\pi\sigma}$ – R dependence plotted in figure 17.

The increase of the intensities of the shoulder A′ and the line A″ for the incoherent phase BN [93] supports the proposed polaron mechanism because density of the defect boron sites in a thin film of the metastable incoherent phase BN is expected to be increased. In fact, Wada and Yamashita [94] have shown that small variations in substrate temperature, laser fluence, ion fluence, and gas pressure will produce BN thin films of hexagonal or incoherent structure. The high sensitivity of the line shape of the π-resonance to short-range order effects was demonstrated in the work [95]. Thus, the line shape of the π-resonance gives a "fingerprint" for recognizing the defect sites.

The existence of the interlayer electronic states for h-BN was theoretically predicted by Catellani and co-workers [96] and Park *et.al.* [97]. The state (labeled Γ_1^+ [96]) at the Γ point is estimated ~0.5 eV above the conduction band minimum at M point and could be responsible for the high-energy shoulder of the prominent B 1s → π^* resonance.

The intense X-ray emission from the $B1s^{-1}2p(\pi)$ excitation makes it possible to assign it as a quasi-stationary defect centers which demonstrates the close resemblance to donor highly localized DX center, for which the dominance of the lattice distortions is established [98].

By comparing the B K-shell excitations in h-BN and carbon allotropes, similar structural changes in the graphite lattice accompanying the low-lying $C1s^{-1}2p(\pi)$ excitation are observed. But, it is very important, that for graphite, there are additional difficulties with respect to strong π − π interaction within a basal plane because the relevant π-MO is more delocalized in a basal plane. However, resemblance of the FWHM ≈ 0.20 eV of the $C1s^{-1}\pi^*$ excitation in graphite [99] could be mentioned. Recent studies of X – ray absorption in single-wall-carbon-nanotubes (SWCNT) give the FWHM of the $C 1s^{-1}\pi^*$ peak equal to 0.32 eV [100]. Intriguing similarity between the fine structures of the peaks in h-BN (Figure 20) and SWCNT attracts high attention. The line shape analysis of the $C 1s^{-1}\pi^*$ peak in SWCNT shows a series of four adjacent peaks with increasing width and spacing (+0.20 eV, +0.35 eV, +0.40 eV) observed exclusively for SWCNT [100]. The characteristic fine structure is assigned to unoccupied van Hove's singularities in bulk SWCNT samples [100].

Thus, the origin of the π^*-band asymmetry in B 1s absorption in h-BN is a very intriguing problem and needs further experimental and theoretical investigations. Below we will discuss appearance of new phenomena in X-ray optics in its vicinity.

NITROGEN EXCITATIONS

The N K-shell absorption spectrum of h-BN crystal [101] is plotted in figure 18 There are three intense absorption line at 401.4 eV (labeled A), 408.3

eV (B) and 416 eV (C). Taking into account the spectral distribution of oscillator strength for N 1s transitions in h-BN (figure 23) computed by using the QA approach under the same approximations outlined before the low-lying absorption line A is assigned to $\pi^*(N1s^{-1}2p_z a_2'')$ resonance and the lines B and C to the first and the second $\sigma^*(N1s^{-1}e'(2p_{x,y}))$ resonances, respectively.

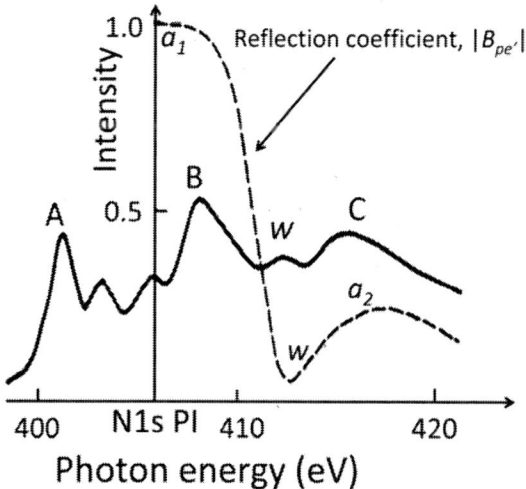

Figure 22. Experimental N K shell absorption spectrum of h-BN taken from the work [101] (solid line). Computed spectral variations of the reflection coefficient $\rho_{pe'} = |B|_{pe'}$ for the N 1s photoelectrons are also plotted (dashed line) [60].

In contrast to B K excitations the N K excitations in h-BN are more delocalized. It is shown in the work [58] the description of the N K-shell excitations is not sufficient when the scattering of the photoelectron at the nearest neighbors [N*$_{1s}$B$_3$] only is taken into account. The second (nitrogen) neighbors [N*$_{1s}$B$_3$N$_6$] exerts a noticeable influence on spectral distribution of oscillator strengths for N 1s $\to \sigma^*$ transitions. The strong difference in localization of B and N 1s excitations is caused by the essential scattering on the second neighbors and the resonance effects within the interstitial region located between the triangle of boron atoms and hexagon of nitrogen atoms.

To understand appearance of the two σ^* shape resonances in spectral distribution of N K shell excitations special emphasis is put on interference of the primary photoelectron with the scattered electrons and the spatial localization of the interferential picture [60]. In contrast to B K excitations N K ones are localized in the cluster composed from two coordination shells. As a consequence of the changes in spatial localization the σ^* shape resonances cannot be attributed to the interatomic B – N distance and depend on the B – N and N – N distances in the cluster. The dependence is much more complicated. To illustrate the complicity spectral dependence of the reflection coefficient for the photoelectron ejected from N 1s level in the [N*$_{1s}$B$_3$N$_6$] cluster is shown in figure 22. Strong spectral variations of the coefficient $\rho_{pe'} = |B|_{pe'}$ are caused by the resonance effects inside the surrounding atoms. These effects can be described as a series of photoelectron reflections from the triangle of the nearest B atoms and the next hexagon of N atoms [101, 102]. In the case eq. (17) is used. The calculations show the appearance of the window in transparency of the N 1s photoelectron through the coordination shells and the regions of abnormal reflections. (In figure 22 the window is labeled via w and the regions via a_1 and a_2). As a consequence of the interferential effects in the surroundings, it as an indivisible subsystem act on nitrogen excitations motivating the appearance of two σ^* resonance features and the strong violation of the smooth correlation between the splitting energies $\Delta_{\pi 1\sigma}(\Delta_{\pi 2\sigma})$ and bonds lengths.

For a molecular simulation of the N K-shell excitations in the crystal the use of complex molecules such as borazine B$_3$N$_3$H$_6$ is required due to the interferential effects. Comparison of the N K-shell absorption spectra of h-BN and planar B$_3$N$_3$H$_6$ molecule [57] supports the importance of the effects. However, the interferential effects in B$_3$N$_3$H$_6$ are obviously weaker than in h-BN. Their weakness explains the decrease of the σ^* resonances relative to the π^* resonance. A more reasonable molecular simulation can be obtained with the help of the cyclic and planar cyanuric acid molecule (CNOH)$_3$ in which the first and second surrounding shells are built up from a pair of nearest carbon atoms and pairs of neighboring N and O atoms. By comparing the experimental N K-shell absorption spectra of (CNOH)$_3$ and B$_3$N$_3$H$_6$ with h-BN a close resemblance of the near N K edge absorption structures is clearly seen, especially for the high energy region dominated by the interferential effects. The intensity of the low-energy π^*-resonance varies substantially as the corresponding π^*-MO is partially filled due to the presence of N – H bonds [61].

The investigations of K-shell excitations in h-BN demonstrate the advantages of their treatment basing on molecular orbitals and reveal a central role of

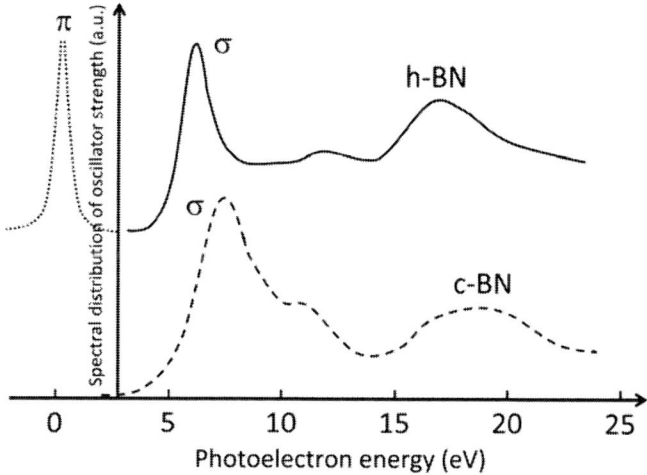

Figure 23. Calculated spectral distribution of oscillator strength for transitions from N 1s level into σ* states in h-BN [58]. The $1\sigma^*$ and $2\sigma^*$ resonances are located at 6 eV and 17 eV. For comparison spectral distribution of oscillator strength for transitions from N 1s level into σ^* states in c-BN crystals is shown too. π^* resonance in h-BN is shown as a dotted line.

i. atomic B and N $1s^{-1}$ 2p resonances in formation of the main resonance features,
ii. splitting of the atomic core excitations by an anisotropic crystal field into single (π) and double degenerated (σ) components due to the main changes in the spectral distribution of oscillator strength for X-ray transitions near the B and N K edges,
iii. change in spatial localization of B and N K-shell excitations.

The different spatial localization of the K excitations is originated by the interference of scattered photoelectron waves. The B K shell excitations are essentially localized within the nearest neighbors. As a result the excitations are tightly related with chemical bonding. In contrast, N K shell excitations are more delocalized and their relationships with chemical bonds are much more complicated.

Chapter 5

OPTICS OF HEXAGONAL BN

OPTICS OF ANISOTROPIC CRYSTALS

In an isotropic media the propagation characteristics of electromagnetic waves are independent on their propagation direction. Generally this means that there is no direction within such a medium which is any different from any other. In an isotropic medium the electric displacement vector **D** and its associated electric field vector **E** are parallel and one can write

$$\mathbf{D} = \varepsilon \varepsilon_0 \mathbf{E} \tag{28}$$

where ε is the scalar dielectric constant, which in the general case is a function of frequency. This is equivalent to saying that the polarization vector **P** of the media induced by the field and the field itself are parallel

$$\mathbf{P} = \chi \varepsilon_0 \mathbf{E} \tag{29}$$

where χ is the scalar susceptibility.

In an anisotropic medium **D** and **E** are no longer necessarily parallel and one can write

$$\mathbf{D} = \varepsilon_{i,j} \mathbf{E} \tag{30}$$

where $\varepsilon_{i,j}$ is the dielectric tensor of rank 2, which in matrix form referred to three arbitrary orthogonal axes is

$$\varepsilon_{i,j} = \begin{pmatrix} \varepsilon_{x,x} & \varepsilon_{x,y} & \varepsilon_{x,z} \\ \varepsilon_{y,x} & \varepsilon_{y,y} & \varepsilon_{y,z} \\ \varepsilon_{z,x} & \varepsilon_{z,y} & \varepsilon_{z,z} \end{pmatrix} \qquad (31)$$

By making the appropriate choice of axes the dielectric tensor can be diagonalized. With this choice of axes Eq. (31) in matrix form becomes

$$\begin{pmatrix} D_x \\ D_y \\ D_z \end{pmatrix} = \varepsilon_0 \begin{pmatrix} \varepsilon_x & 0 & 0 \\ 0 & \varepsilon_y & 0 \\ 0 & 0 & \varepsilon_z \end{pmatrix} \times \begin{pmatrix} E_x \\ E_y \\ E_z \end{pmatrix} \qquad (32)$$

where $\varepsilon_x, \varepsilon_y, \varepsilon_z$ are called the principal dielectric constants. The directions x, y and z are in this case known as the principal axes of the medium. Note that these axes are not necessarily orthogonal.

Alternatively, we can describe the anisotropic character of the medium with the aid of the susceptibility tensor $\chi_{i,j}$

$$\mathbf{P} = \chi_{i,j} \varepsilon_0 \mathbf{E} \qquad (33)$$

Since $\mathbf{D} = \varepsilon_0 \mathbf{E} + \mathbf{P}$, it is clear that in a principal coordinate system $\varepsilon_x = 1 + \chi_x$; etc.

Let us assume that a monochromatic plane wave propagates through an anisotropic medium. The direction of **D** species the direction of polarization of this wave. The wave vector **k** of this plane wave is normal to the wavefront (the plane where the phase of the wave is everywhere equal) and has magnitude $|k| = \omega/\upsilon$, where υ is the phase velocity of the wave, which is related to the velocity of light in vacuum, **c**, by the refractive index n experienced by the wave according to $\upsilon = c/n$. We stress that **D** and **E** are now related by a tensor operation. Then in the principal coordinate system the following equation can be written:

$$D_x = \frac{-K_x P^2}{v^2 - v_x^2} \tag{34}$$

Analogous equation can be written for y and z components. Respectively K_x, K_y, K_z are the three orthogonal components of the wave vector and

$$v_x = \frac{c}{\varepsilon_x^{1/2}}, \quad v_y = \frac{c}{\varepsilon_y^{1/2}}, \quad v_z = \frac{c}{\varepsilon_z^{1/2}}$$

are called the principal phase velocities of the crystal. Then one can obtain the equation, which is called Fresnel equations [103]:

$$\frac{K_x}{v^2 - v_x^2} + \frac{K_y}{v^2 - v_y^2} + \frac{K_z}{v^2 - v_z^2} = 0 \tag{35}$$

$$\frac{K_x n_x^2}{n^2 - n_x^2} + \frac{K_y n_y^2}{n^2 - n_y^2} + \frac{K_z n_z^2}{n^2 - n_z^2} = 0 \tag{36}$$

where $n_x = \varepsilon_x^{1/2}$, $n_y = \varepsilon_y^{1/2}$, $n_z = \varepsilon_z^{1/2}$ are called the principal refractive indices of the crystal. The equations (35) and (36) are quadratic in c^2 and n^2, respectively. Thus in general there are two possible solutions v_1; v_2 and n_1; n_2, respectively, for the phase velocity and refractive index of a monochromatic wave propagating through an anisotropic medium with wave vector **k**. However, when **k** lies in certain specific directions both roots of Eqs. (35) and (36) become equal. These special directions within the crystal are called optic axes. One can show that the two solutions of Eqs. (35) and (36) correspond to two different possible linear polarizations of a wave propagating with wave vector **k** and that these two solutions have mutually orthogonal polarization.

Let us look now on the **D**, **E** and **H** vectors. The electric displacement vector **D** and wave vector **k** are, by definition, mutually perpendicular. The values of **D**, **E** and **H** are constant over the phase front of a plane wave. All of this provides certain angular relationships between them. **H** is normal to both **D**, **E** and **k**, and the latter three are coplanar. **D** and **E** make an angle α with one another where

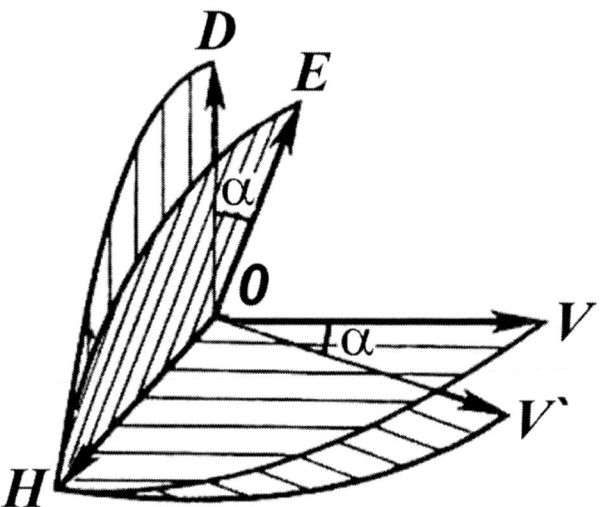

Figure 24. Angular relationships between **D**, **E**, **H** vectors.

$$\alpha = \arccos\left(\frac{ED}{|E||D|}\right) \quad (37)$$

These angular relationships are illustrated in Figure 24.

The direction of the Pointing vector $S = E \times H$ is, by definition, perpendicular to both **E** and **H** and defines the direction of energy flow within the medium. We identify it as the direction of the ray of familiar geometric optics. Whereas in isotropic media the ray is always parallel to the wave vector, and is therefore perpendicular to the wavefront, this is no longer so in anisotropic media, except for propagation along one of the principal axes.

One can conclude that transverse electromagnetic plane waves can propagate through anisotropic media, but for propagation in a general direction two distinct allowed linear polarizations specified by the direction of **D** can exist for the wave. These two allowed polarizations are orthogonal and the wave propagates with a phase velocity (the velocity of the surface of constant phase - the wavefront), which depends on which of these two polarizations it has. Clearly, a wave of arbitrary polarization which enters such an anisotropic medium will not in general correspond to one of the allowed polarizations, and will therefore be resolved into two linearly polarized components polarized along the allowed directions. Each component propagates with a different phase velocity.

Dependence of a dielectric permittivity ε and, thus, refractive index $n (n = \sqrt{\varepsilon})$ from an electric field direction can be presented graphically. A geometric figure that shows the index of refraction and vibration direction for light passing in any direction through a material is called an optical indicatrix (figure 25). The indicatrix is constructed by plotting indices of refraction as radii parallel to the vibration direction of the light. If the indices of refraction for all possible light rays are plotted in a similar way, the surface of the indicatrix is defined. The shape of the indicatrix depends on crystal symmetry. The indicatrix, wave-normal, is an ellipsoid with the equation [104]:

$$\frac{x^2}{n_x^2} + \frac{y^2}{n_y^2} + \frac{z^2}{n_z^2} = 1 \qquad (38)$$

where n_x, n_y and n_z are refraction indexes along the principle axis. In a case of optically isotropic cubic crystals ε does not depend on a direction and optical indicatrix is sphere with radius $r = n = \sqrt{\varepsilon}$. In crystals belonging to the tetragonal, hexagonal and trigonal crystal systems the crystal symmetry requires that $n_x = n_y$ and the indicatrix reduces to an ellipsoid of revolution. In this case there is only one optic axis, oriented along the axis of highest symmetry of the crystal, the z axis (or c axis). Such crystals are known as uniaxial anisotropic crystals. Crystals belonging to the less symmetric orthorhombic, monoclinic and triclinic crystal systems has an indicatrix, which is an ellipsoid with three principle axes, and they are called biaxial anisotropic crystals.

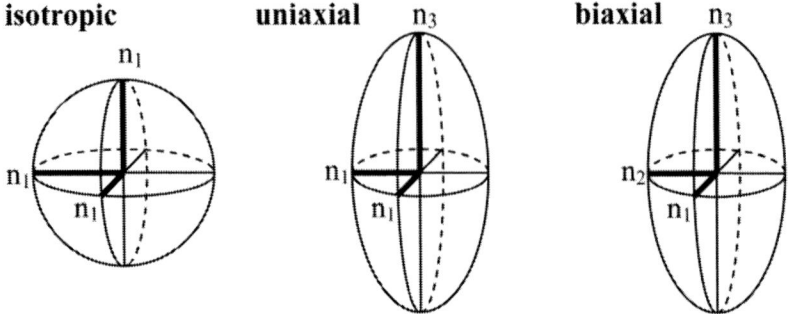

Figure 25. Optical indicatrixes of isotropic, uniaxial and biaxial crystals.

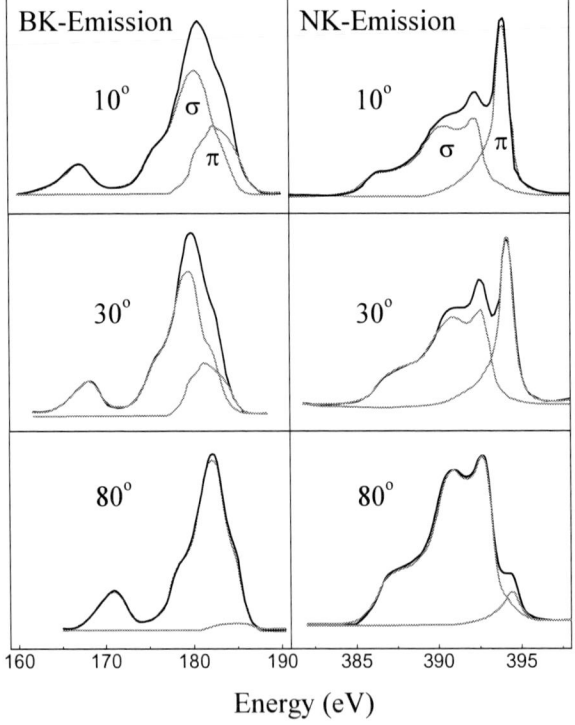

Figure 26. K-emission spectra of boron and nitrogen in h-BN for three different take-off angles obtained by Tegeler and co-authors [105].

The equation of the uniaxial indicatrix is:

$$\frac{(x^2+y^2)}{n_0^2} + \frac{z^2}{n_0^2} = 1 \qquad (39)$$

where n_0 is the index of refraction experienced by waves polarized perpendicular to the optic axis, called ordinary or O - waves; n_e is the index of refraction experienced by waves polarized parallel to the optic axis, called extraordinary or E - waves. If $n_e > n_0$ the indicatrix is a prolate ellipsoid of revolution and such a crystal is said to be positive uniaxial. If $n_e < n_0$ the indicatrix is an oblate ellipsoid of revolution and the crystal is said to be negative uniaxial. Because uniaxial crystals have indicatrixes which are circularly symmetric about the z (optic) axis, their optical properties depend only on the polar angle, that the wave vector **k** makes with the optic axis and not on the azimuthal orientation of **k** relative to the x and y axes.

Chapter 6

X-RAY RADIATION INTERACTION WITH HEXAGONAL BN CRYSTAL

In the articles [105, 106] angle-resolved measurements of emission spectra of h-BN were presented. The K-emission spectra of boron and nitrogen in h-BN for three different take-off angles, obtained by Tegeler and co-authors [105] are shown in figure 26. The spectra were measured at take-off angles 10°, 30° and 80°. A separation of either emission band into π- and σ-components are shown in the figure, too. As follows from the figure the intensity distribution of the emission band for any take-off angle can be represented as a superposition of the π and σ subbands. The subbands obtained are shown as dashed lines in figure 26.

As can be seen in figure 26 the shape of the BK-emission spectrum only slightly depends on the take-off angle in contrast to NK-emission spectrum. The strong dependence of the shape of the NK-emission spectrum upon the take-off angle is very pronounced but only in the π- subband. The effect is less marked in the case of BK – emission band because the intensity of the π-subband relative to that σ -subband is smaller for boron than for nitrogen (figure 8).

In the figure 27 the angular dependencies of BK and NK – absorption spectra [52, 108] are presented. We underline that the dependencies are conditioned by angular dependencies of the measured X-ray reflection spectra. The spectra were measured at incident angle 90° [59] and 45° [108] degrees. As seen from the figure 27 the angular dependence of the absorption spectra is again pronounced in the spectral features connected with transitions to π states.

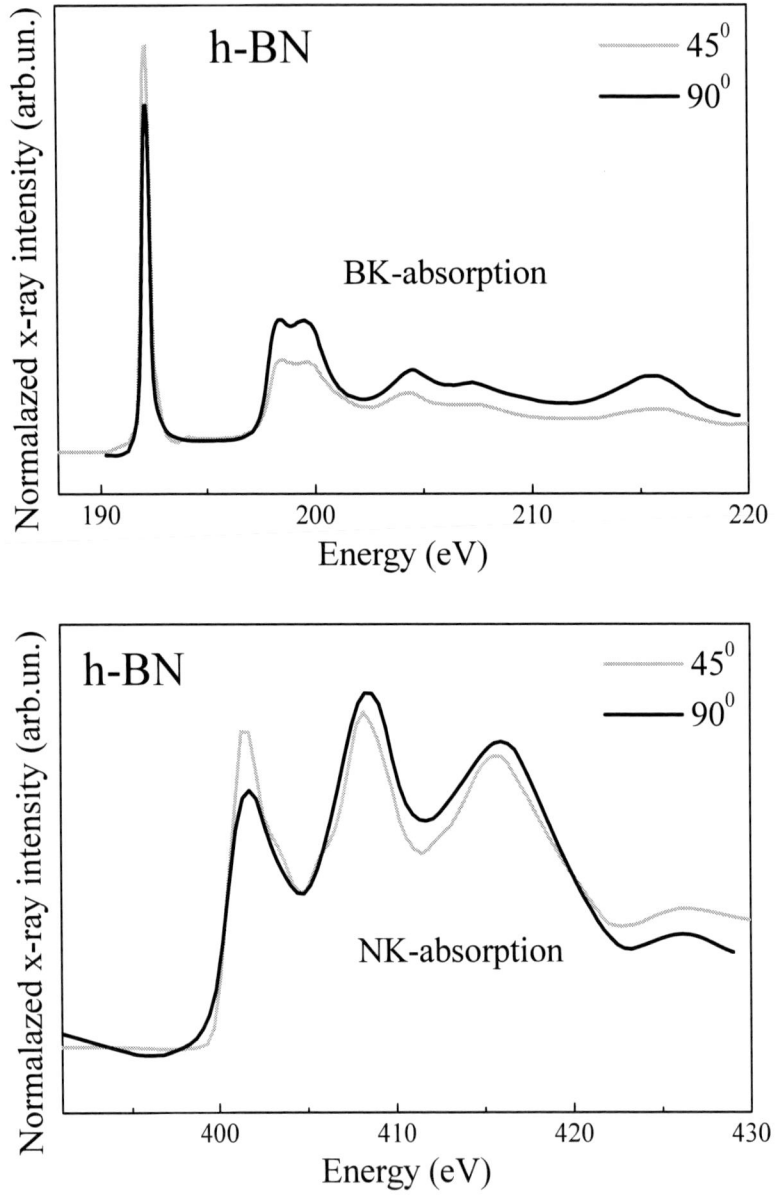

Figure 27. The angular dependencies of BK and NK – photoyield spectra. The spectra were measured at incident angle 90° in [59] and 45° by Kawaguchi and co-authors [108].

ORIENTATION AND POLARIZATION DEPENDENCE OF REFLECTED RADIATION

To understand the peculiarities of the interaction of h-BN with x-ray radiation from spatial crystal point of view let us analyze the total external reflection of x-rays from h-BN different oriented about the electric field vector **E**. Different orientations of the crystal h-BN is plotted in the Figure 28.

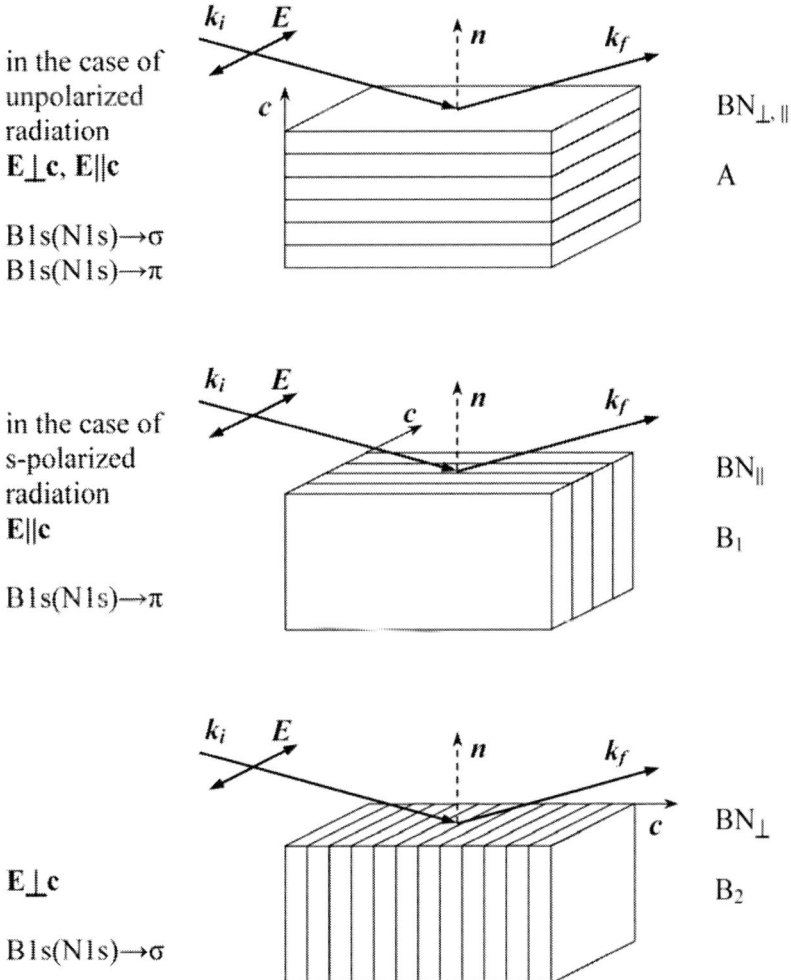

Figure 28. Different orientations of the crystal h-BN about the electric field vector **E**. A, B_1 and B_2 label the crystal orientation.

X-ray total external reflection (specular reflection) is an integrated process and is defined by both the photoabsorption process and the susceptibility of the material. The primary photon $\hbar\omega$ impinges on a sample and photoionization takes place; locally the photon is absorbed and an electron is excited into empty states. Besides, the interaction of the photon $\hbar\omega$ with sample leads to the polarization of the substance. As a result the system comes back into the undisturbed state when the core-hole is filled and the photon is emitted. The intensity of the "reflected" photons is detected. The reflection spectra R(E) exhibit "edges" due to the elements present in the specimen and are very sensitive to the nature of the absorbing atoms, their chemical state and local coordination environment. Because the angular dependence of a penetration depth of the reflected beams, X-ray reflection spectroscopy is in-depth characterization tool of the local atomic structure of materials.

Hexagonal BN is a one-axis crystal. In the case of a one-axis crystal the dielectric function tensor is always a diagonal matrix if one of the axes of a system of coordinates is aligned with the optical axis (**c** axis) of the crystal. In this case $\varepsilon_{xx} = \varepsilon_{yy}$ is valid. In the case of the polarized radiation the problem is simplified. According to [109] the Fresnel equations for s- and p-linearly polarized radiation are given by:

$$R_s = \left| \frac{\sin\theta - \sqrt{\varepsilon_{xx} - \cos^2\theta}}{\sin\theta + \sqrt{\varepsilon_{xx} - \cos^2\theta}} \right|^2 \quad (40a)$$

$$R_p = \left| \frac{\sqrt{\varepsilon_{yy}\varepsilon_{zz}}\sin\theta - \sqrt{\varepsilon_{zz} - \cos^2\theta}}{\sqrt{\varepsilon_{yy}\varepsilon_{zz}}\sin\theta + \sqrt{\varepsilon_{zz} - \cos^2\theta}} \right|^2, \quad (40b)$$

where the axes X and y of a system of coordinates are located in the plane of the reflecting surface of the crystal and the axes y and z are located in the plane of the scattering. One can see from (40a) that with the use of p-polarized radiation the reflection coefficient depends on two components of the dielectric function matrix $\varepsilon_{yy} = \varepsilon_\perp$ and $\varepsilon_{zz} = \varepsilon_\|$, that is it depends on two components of electric field vector **E** (perpendicular and parallel to the **c** axis of the crystal). When using s-polarized radiation the reflection coefficient is defined by the component of the dielectric function matrix for the direction of the electric field vector **E**, which is perpendicular to the **c** axis of the crystal ($\varepsilon_{xx} = \varepsilon_{\parallel}$). By this means we assume that the axes X and y of a coordinate system are located in the plane of the reflecting crystal surface which is perpendicular or parallel to the **c** axis of the crystal. By turning the crystal the $\mathbf{E} \perp \mathbf{c}$ and $\mathbf{E} \parallel \mathbf{c}$ alignments are realized and the orientation dependence of the reflection and absorption spectra can be investigated. The greatest effects would be expected when polarized radiation is used, especially in the case of s-polarized radiation; however, these effects have to exist in the reflection spectra of unpolarized radiation.

Using the dipole approximation, the transition probability of the electron from the core levels to the free states of the conduction band is given by

$$W \sim \left| \int \psi_j^* z \psi_i d\tau \right|^2 \qquad (41)$$

where the Z axis is parallel to the **c** axis of the crystal and ψ_i and ψ_j are wavefunctions of the initial and final states, respectively, and $d\tau$ is a volume element. Because K-shell excitations are viewed, the initial state of the transitions is the 1s level of the B (N) atom, which has a spherical even wavefunction and always is even relative to the reflection in the layer plane. As can be seen from (41), only a transition probability to an odd state is not equal to zero. When $\mathbf{E} \perp \mathbf{c}$ alignment is realized only transitions to the final states with even wavefunction ψ_j are allowed. So according to the dipole selection rules with regard to the crystal symmetry, the absorption in h-BN must originate from transitions from B 1s (N 1s) to σ states in the case $\mathbf{E} \perp \mathbf{c}$ and from B 1s (N 1s) to π states in the case of $\mathbf{E} \parallel \mathbf{c}$.

Using polarized radiation one would expect the polarization dependence of the K-shell x-ray reflection near edge structure in h-BN to be a representation of polarization dependence of the absorption spectra. When unpolarized radiation is used and taking into account the cross nature of electromagnetic waves ($\mathbf{E} \perp \mathbf{k}_i$, where $\mathbf{k}_i$ is the photon wavevector) one can change the ratio of components $\mathbf{E} \perp \mathbf{c}$ and $\mathbf{E} \parallel \mathbf{c}$ by turning the crystal. By means of this rotation, one can change the contributions of different transitions to the B K- (N K-) absorption spectra.

Different orientations of the textured polycrystalline h-BN crystal about the electric field vector **E** are shown in figure 28. As follows from the figures when s-polarized radiation is used the absorption spectrum is largely originated from transitions $B1s(N1s) \rightarrow \pi$ states in the case of the orientation B_1 (**E** $\parallel$ **c**) and from transitions $B1s(N1s) \rightarrow \sigma$ states in the case of the orientation B_2 (**E** $\perp$ **c**). When unpolarized radiation is used the contribution of each component (**E** $\perp$ **c** and **E** $\parallel$ **c**) to the overall reflection process is analyzed. For orientation B_2 the electric field vectors are always aligned in the plane parallel to the basal plane and so geometry like **E** $\perp$ **c** is realized in this case. For orientation A the vectors **E** can be arranged within (**E** $\perp$ **c**) as well as perpendicular (**E** $\parallel$ **c**) to the basal plane.

B AND N K-REFLECTION SPECTRA OF HEXAGONAL BN

The experimental studies of the interaction of x-ray radiation with h-BN [110-113] show a strong orientation dependence of x-ray reflectivity with respect to the electric field vector. Figure 29 shows the B K-reflection spectra obtained for grazing incidence angles 4° and 8° for the crystal orientations B_1 and B_2 using s-polarized radiation [110]. The B K-reflection spectra obtained for crystal orientations A and B_2 using unpolarized [110, 111] radiation are plotted in the figure 30.

Analysis of the reflection spectra fine structure (figures 29, 30) indicates that the reflection spectra for different orientations are similar in the number of essential features and their energy positions for investigated angles. The shape of the reflection spectra for small incidence angle 4° (smaller than critical angle of total external reflection) depends only slightly on the crystal orientation. As this takes place the ratio between magnitudes of peaks M and N is found to be larger for orientation B_1 as compared with orientation B_2 (figure 29) and for orientation A as compared with orientation B_2 (figure 30). The growth of the incidence angle to 8° leads to a significant increase of the intensity of peak M in the case of orientations B_1 and A. Strong intensity dependence due to crystal orientation exists for peak M for this angle. Recall that exactly in the case of crystal orientations B_1 and A the absorption spectrum is essentially originated from the transitions $B1s \rightarrow \pi$ states.

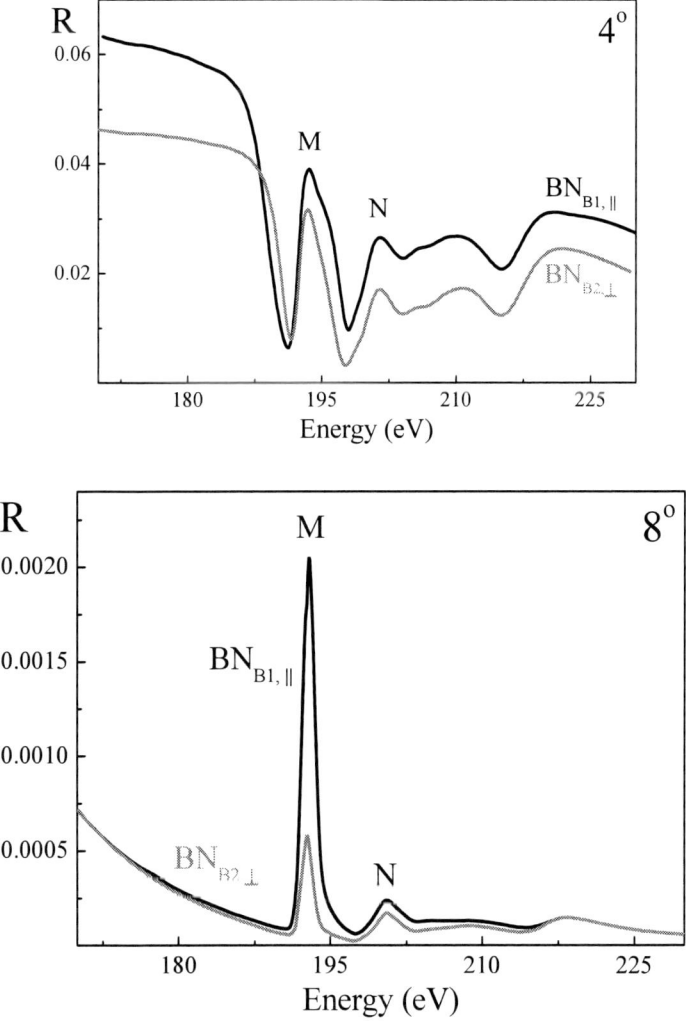

Figure 29. B K-reflection spectra of h-BN obtained for two grazing incidence angles for crystal orientations B_1 and B_2 using s-polarized radiation. The black curve denotes the data for the orientation B_1 ($BN_{B1, \parallel}$) and the grey curve denotes the data for orientation B_2 ($BN_{B2,\perp}$) (a) $\theta = 4°$ (b) $8°$ [110].

Figure 31 shows the N K-reflection spectra obtained for grazing incidence angles 3° and 4° for crystal orientations B_1 and B_2. The orientation dependence of feature M in the reflection spectra for both angles has engaged our attention. A higher magnitude of peak M is detected for orientation B_1. In this case transitions $B1s \rightarrow \pi$ states are dominant. The orientation dependence of feature O in the reflection spectra obtained for incidence angle 4° is noteworthy too. A higher magnitude of peak O is observed for orientation B_2. In this case transitions $B1s \rightarrow \sigma$ states are dominant.

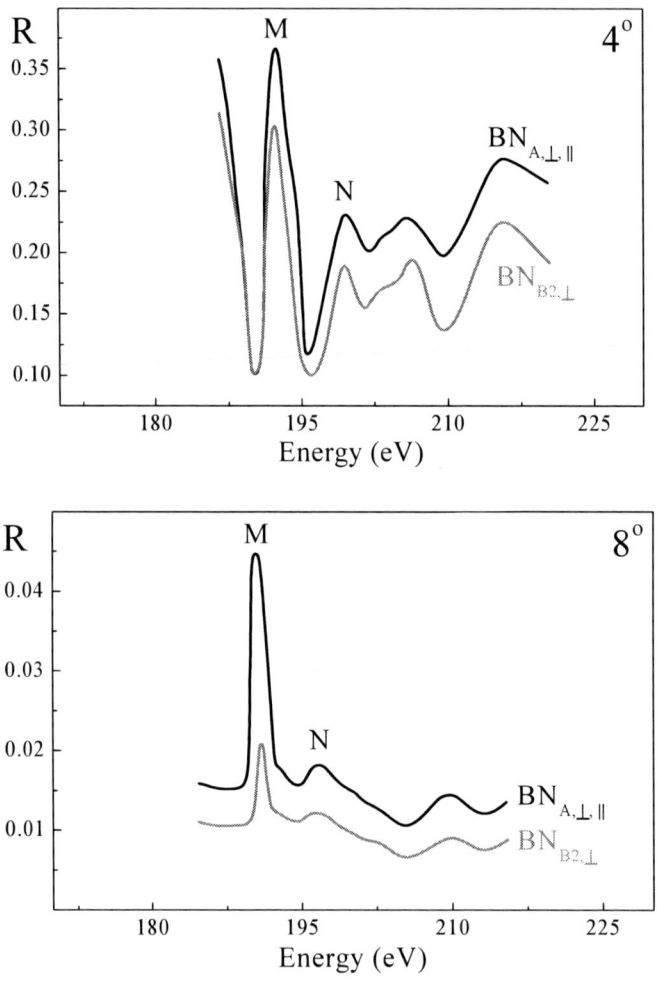

Figure 30. B K-reflection spectra of h-BN obtained for grazing incidence angles 4° and 8° for crystal orientations A and B_2 with use of unpolarized radiation. The black curve denotes the data for orientation A ($BN_{A, \perp, \parallel}$) and the grey curve denotes the data for orientation B_2 ($BN_{B2, \perp}$) (a) $\theta = 4°$, (b) 8° [110, 111].

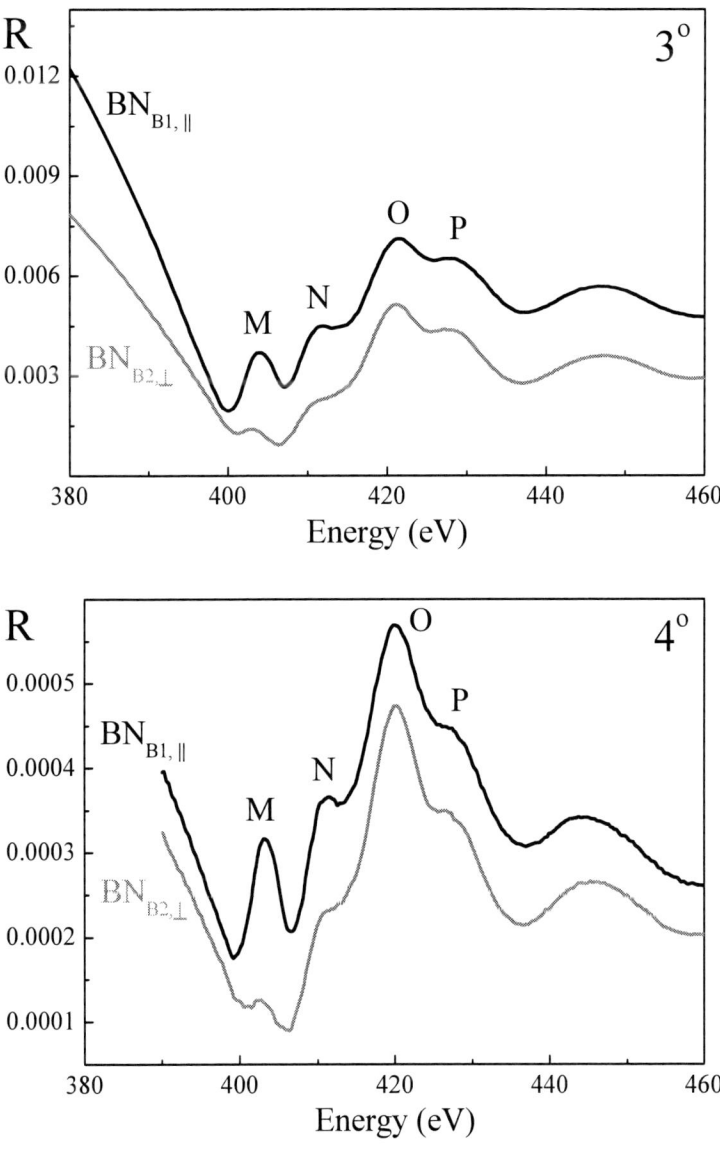

Figure 31. N K-reflection spectra of h-BN obtained for grazing incidence angles for crystal orientations B_1 and B_2 using s-polarized radiation. The black curve denotes the data for the orientation B_1 ($BN_{B1,\parallel}$) and the grey curve denotes the data for orientation B_2 ($BN_{B2,\perp}$) (a) $\theta = 3°$ (b) $4°$ [110].

The joint examination of B and N K-reflection spectra for different orientations allows definite conclusions. Peak M defines significantly the orientation dependence of B K-reflection spectra. It is the intensity of peak M, which is the most sensitive to the changes of the incidence angle and to the crystal orientation. A completely different type of situation occurs in the N K reflection spectra. In this case all peaks are involved equally in formation of the fine structure. The orientation dependence is described by peaks M and O. Hence the joint analysis of B and NK-reflection spectra points to the different evolution of the formation of K-reflection spectra near B- and N-edges.

KRAMERS-KRONIG ANALYSIS

Because of the low polarizibility of substances, the dielectric constant in the x-ray range can be conveniently represented in the form $\varepsilon = 1 - \varepsilon_1 + i\varepsilon_2$. The value ε_1 is usually positive.

Within the framework of the Fresnel model, i.e., for a perfectly smooth isotropic surface of a substance, the reflection amplitude r of the s- and p-polarized radiation can be written as:

$$r_s = \frac{\sin\theta - \sqrt{\varepsilon - \cos^2\theta}}{\sin\theta + \sqrt{\varepsilon - \cos^2\theta}} \tag{42a}$$

$$r_p = \frac{\varepsilon\sin\theta - \sqrt{\varepsilon - \cos^2\theta}}{\varepsilon\sin\theta + \sqrt{\varepsilon - \cos^2\theta}} \tag{42b}$$

where θ is the angle of grazing incidence.

In the region of small grazing angles one it is possible not to consider polarization and to use the Fresnel formula as:

$$r = \frac{a - ib - \sin\theta}{a - ib + \sin\theta} \tag{43}$$

Where

$$\begin{cases} a^2 - b^2 = n^2 - k^2 - \cos^2\theta \\ ab = nk \end{cases}$$

We recall that $\widetilde{\varepsilon} = \widetilde{n}^2$, $\widetilde{n} = n + ik$, $\begin{cases} 1-\varepsilon_1 = n^2 - k^2 \\ \varepsilon_2 = nk \end{cases}$, and $\mu = 4\pi k/\lambda$, where μ is a linear absorption coefficient. In the experiment, the reflection coefficient $R = |r|^2$ for intensities is measured. The complex reflection coefficient r can be expressed in terms of its modulus $|r| = R^{1/2}$ and its phase shift ψ appearing after reflection of the wave at the surface of the medium:

$$\sqrt{R}e^{i\psi} = \frac{a - ib - \sin\theta}{a - ib + \sin\theta} \tag{44}$$

One can deduce from the relation (43):

$$\begin{cases} a = \dfrac{(1-R)\sin\theta}{(1+R) - 2\sqrt{R}\cos\theta} \\ b = \dfrac{-2\sqrt{R}\sin\theta\sin\psi}{(1+R) - 2\sqrt{R}\cos\psi} \end{cases}$$

The value of ψ can be determined using the Kramers–Kronig dispersion relation:

$$\psi(E_0) = \frac{E_0}{\pi} v.p. \int_0^\infty \frac{\ln R(E)}{E^2 - E_0^2} dE \tag{45}$$

For the application of the Kramers–Kronig analysis one measures the absolute reflectivity $R(E_0)$ at a fixed grazing angle as a function of the photon energy. The use of the Kramers–Kronig integral can involve difficulties, which are considered in detail in papers [144-151]. It is known [144-147] that generally the integral in a form (45) ambiguous defines dependence between imaginary and real parts of function, i.e. additional arbitrary constants along with own dispersive integral are possible. But in the case under consideration, the phase shift ψ, appearing after reflection of the wave at the surface of the medium, changes in limits of $-\pi \leq \psi \leq 0$. Then, according to analysis of the appearance of additional constants in the (45) carried out in [144-147], the phase shift can be uniquely deduced from reflection coefficients R using (43). Uncertainty in the determination of the phase shift ψ in this case results from the limited energy range where the reflection coefficient can be measured. It is convenient to rewrite the integral (43) as:

$$\psi(E_0) = \frac{E_0}{\pi}\int_0^{E_1} f(R,E)dE + \frac{E_0}{\pi}\int_{E_1}^{E_2} f(R,E)dE + \frac{E_0}{\pi}\int_{E_2}^{\infty} f(R,E)dE, \quad (46)$$

where the energy region (E_1, E_2) corresponds to the domain of measurement. Obviously, the necessity of using the extrapolation of the reflection spectrum outside the experimental energy range.

Some interesting consequences of the dispersion relation (22) are best displayed by carrying out it to the form:

$$\psi(E_0) = \frac{1}{2\pi} v.p. \int_0^{\infty} \ln\left|\frac{E+E_0}{E-E_0}\right| \frac{d(\ln R(E))}{dE} dE \quad (47)$$

One can see from equation (47) that main contribution to the value of phase shift $\psi(E_0)$ gives the spectral regions in the neighborhood of E = E_0 (logarithm $\ln\left|\frac{E+E_0}{E-E_0}\right|$ will have strong peak in the neighborhood of E = E_0) and spectral regions where the function R(E) changes rapidly that leads to large value of the derivative.

With reference to ultrasoft X-ray spectra it means: the extrapolation to the hard X-ray spectral region where the reflection coefficient typically decreases sharply, and to the UV spectral region where the oscillations of R(E) related to the interband transitions are usually observed, can substantially affect the value of the phase shift ψ. Nevertheless, according to papers [114, 152], one can choose the integration domain is large enough not to affect the physical results. As follows from the papers [114, 152], at small grazing incident angles, the relation $R(E) \sim E^{-4}$ can be used for the extrapolation of the experimental reflection spectrum toward high energies, such that $\theta = \theta_c > 2$. In the case of reflection spectra of heavy elements the $\theta = \theta_c > 3.7$ is more preferable. As for extrapolation of the reflection spectrum to the low-energy region, it was shown in paper [153] that calculated data only slightly depend on the type of extrapolation to this region. For this reason, one can simply drew a smooth curve connecting the experimental spectrum with point R(0)=1

Chapter 7

B AND N K - ABSORPTION SPECTRA OF H-BN

To interpret the effects established in reflection spectra the absorption spectra were calculated from the reflection spectra by means of Kramers–Kronig relations. In the strict sense the applicability of Kramers–Kronig analysis to the anisotropic crystal is not evident. The main problem in this case is connected with taking into account the surface roughness. Because of the composite micro-structure the surface of any crystal side (cut parallel or perpendicular to the **c** axis) will be disrupted by splits after any type of technological polishing. So a prerequisite for the formation of scattering into the substance, in addition to scattering into vacuum, is produced. Up until now no theoretical description of the scattering deep into the material has been available. This is the reason that the surface roughness was not taken into account in the calculations of the absorption spectra. According to a large body of research the relative differences in the calculated spectra will be correct.

The absorption spectra were calculated from the experimental reflection spectra by means of Kramers–Kronig relations. As an example B K-absorption spectra of h-BN calculated from the experimental reflection spectra obtained for grazing incidence angle 4° for crystal orientations A and B_2 using unpolarized radiation are plotted in figure 32.

The main problem in this case is connected with taking into account the surface roughness. Because of the composite micro-structure the surface of any crystal side (cut parallel or perpendicular to the **c** axis) will be disrupted by splits after any type of technological polishing. So a prerequisite for the formation of scattering into the substance, in addition to scattering into vacuum, is produced. Up until now no theoretical description of the scattering deep into the material has been available. This is the reason that the surface roughness was not taken into account in the calculations of the absorption spectra. According to a number of researches the relative differences in the calculated spectra will be correct.

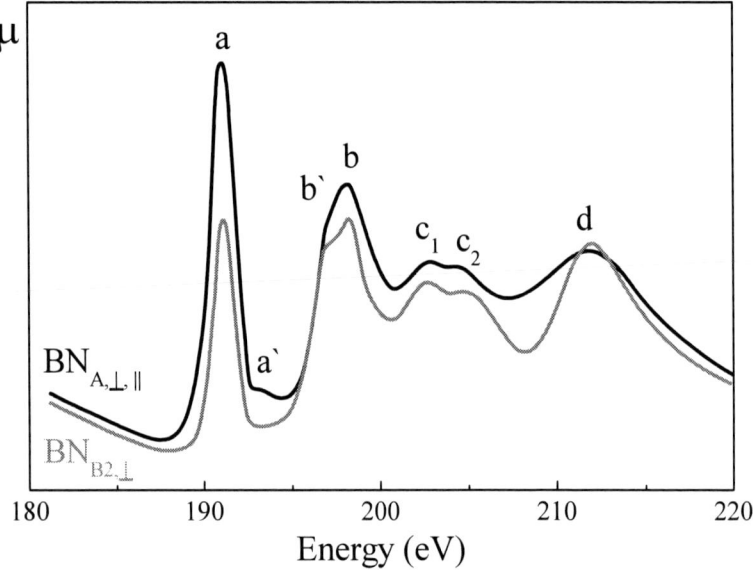

Figure 32. B K-absorption spectra of h-BN calculated from reflection spectra obtained for grazing incidence angle 4° for crystal orientations A and B_2 using unpolarized radiation [110]. The black curve denotes the data for orientation A ($BN_{A, \perp, \parallel}$) and the grey curve denotes the data for orientation B2 ($BN_{B2, \perp}$).

Analysis of the calculated B K-absorption spectra shows that the energy position of all features is independent of the crystal orientation that was used in the calculations whereas the intensity of all features depends on the crystal orientation. One can see that the selective line a dominates in the calculated B K-absorption spectra. The value of selective absorption in peak a is larger in the spectrum obtained for orientation A (figure 28) for which the electric field vector **E** can be arranged both within and perpendicular to the basal plane. The intensity of features b, c. d is larger for the crystal orientation when the alignment $\mathbf{E} \perp \mathbf{c}$ is realized (figure 28, B_2) and transitions B1s → σ states are dominant in this case. The dynamics of the orientation dependences of the calculated BK-absorption near edge structure correlates well with the calculations carried out in the framework of the quasi-atomic approach, presented above. According to these calculations the B K-excitations are essentially localized within the nearest neighbors and the reasonable description of the B 1s excitations can be obtained by using the small B^+N_3-cluster approximation. The π interaction within a basal (xy) plane as well as between the neighboring layers can be neglected in this case. The features a and b in figure 32 can be evidently assigned with the absorption bands A and B in figure 16 (a) associated respectively with the single $B1s \rightarrow 2p_z(a_2'')$ and double degenerated $B1s \rightarrow e'(2p_{x,y})$ transitions. Thus, these spectral features can be attributed to the splitting components of the B 1s → 2p resonance. Essentially, the $2p_z(a_2'')$-splitting component is oriented normally to the basal plane and the $e'(2p_{x,y})$-splitting component is located within a basal plane.

The dynamics of the orientation dependence of feature c and d in the calculated spectra supports the assumption that this features can be also assigned with weak oscillation in spectral distribution of oscillator strength for transitions to $p_{x,y}e'$ states and with electron scattering on neighboring atoms in the basal plane.

One can see that feature b (σ resonance) has double structure. It is significant that the much clearer resolution of this structure (b, b') is observed in the spectra corresponding to the crystal orientation when $\mathbf{E} \perp \mathbf{c}$, that is the transitions to σ states (localized within the basal plane) become dominant. The average splitting value $\Delta_{\pi\sigma} = \frac{1}{2}(\Delta_{\pi\sigma_1} + \Delta_{\pi\sigma_2})$ for the σ resonance maxima (structure b, b') is equal to 6.73 eV and correlates well with the experimental dependence of $\Delta_{\pi\sigma}$ on the interatomic distance R extracted from B 1s spectra of h-BN and planar molecules BF_3, BCl_3, BBr_3, $B_3N_3H_6$, and planar BO_3^{3-} complex anions in cubic crystals [59]. The Jahn–Teller and multielectron effects seems to be plausible for the explanation of the appearance of the observed doublet structure. In other words, the doublet structure of feature b can be correlated with the additional splitting of the doubly degenerate $e'(2p_{x,y})$ excitation due to low symmetry distortions of the nearest environment. Notice that the obtained energy splitting $\Delta_{\pi\sigma}$ agrees well with the value $\Delta_{\pi\sigma} = 6.75$ eV taken from the B K-absorption spectrum measured for h-BN in [58]. In addition we stress that the doublet structure could be assigned with the multielectron excitations associated with inner-shell ionization.

Now let us consider shoulder a' which emerges in the vicinity (upward shift ≈2 eV) of selective peak a. Analysis of the orientation dependences shows that shoulder a' is only manifested in the B K-absorption spectra calculated from the reflection spectra obtained in the geometry when transitions B1s→ π states are dominant (orientation A, figure 28; that is, the origin of shoulder a' connects with the transitions to π states. At the same time because peak a is assigned to the single $B1s \to 2p_z(a_2'')$ resonance shoulder a' cannot be assigned to a single-electron transition. According to analysis presented above π-resonance in h-BN can be presented as a sum over the $B1s^{-1}2p\pi$-excitations localized in the planar and pyramidal [BN$_3$] configurations induced by the relaxation near the core-excited atom. For the pyramidal C$_{3v}$ configuration the local atomic deformation takes place. Such a configuration is achieved when the nitrogen atoms are fixed at their regular positions and the π-core-excited B atom is an out-of-plane displacement (of ≈0.16 A) of the core-excited atom resulting in defect boron sites in the crystal. Using the well-known "Z+1" analogue model the core-excited $B^*(1s^{-1}2p\pi^1)N_3$ fragment can be replaced by the pyramidal CN$_3$-one. Then, taking into account the $\Delta_{\pi\sigma}(R_e)$ relationship one may estimate the magnitude of the out-of-plane displacement (see discussion in part 1).

Figure 33 shows N K-absorption spectra calculated from the reflection data. These spectra demonstrate clearly the orientation dependence that was established during the analysis of the B K-absorption spectra, that is the intensity of peak a is greater for the geometry $\mathbf{E} \parallel \mathbf{c}$, and the intensities of peaks b and c are larger for the geometry $\mathbf{E} \perp \mathbf{c}$. One can see that there is clear orientation dependence for feature a in N K-absorption spectra too, but in this case feature a is not dominant in the spectra. The N K-absorption spectrum demonstrates three intense absorption lines (a, b, c) and two broad absorption bands (d and e).

The surprising thing is that the dependence of the energy position of all features and the incidence angle takes place in this case. The spectra $\mu_{4^0}(E)$ are shifted downward with respect to the spectra $\mu_{3^0}(E)$ by ≈2 eV. As this takes place the energy distance between peaks a and b is conserved and that between peaks b and c is decreased in spectra for both orientations. The growth of the incidence angle leads to an intensity redistribution between peaks b and c especially in the case of crystal orientation B_2. All of these have verified the theoretical prediction that the interpretation of the N K-absorption spectra taking into account only the nearest neighbors is not sufficient, as in the case of the BK-absorption spectra. The resonance effects inside surroundings (see, chapter 3, eq. (17)), i.e. inside the nearest and next nearest coordination shells, dominate in this case. Just the second coordination shell formed by six atoms of N exerts a noticeable influence on the σ^* excitations while the effects of the third and other distant shells on the $\mu(E)$ are weak. In QA approach peak a is assigned to is assigned to $\pi^*(N 1s^{-1} 2p_z a_2'')$ resonance and the lines b and c to the first and the second σ^* $(N 1s^{-1} e'(2p_{x,y}))$ resonances, respectively.

According to these calculations both σ^* resonances (features b and c) are located in the regions of abnormal reflection and shoulder ε falls into the 'window' of transparency [54, 55, 57]. Analysis of the angular and orientation dependences of the NK-absorption spectra (figure 29) testifies to the fact that with the growth of the incidence angle the energy position of shoulder ε remains the same but the energy distance Δ_{b-c} between peaks b and c falls from 8.8 to 7.4 eV. Notice that the calculated value in [58] is $\Delta_{b-c} = 10.04$ eV, and the meaning $\Delta_{b-c} = 7.5$ eV was obtained in [58] from the experimental absorption spectrum. Because the surroundings as an indivisible subsystem act on atomic excitations motivating the appearance of peaks b and c these peaks connect with peculiarities of the electronic structure. So, it would appear reasonable that the change of Δ_{b-c} is caused by variations in the crystalline structure. Really the change of the grazing incident angle leads to the change of the depth of formation of the reflected beam. According to [115] the technological treatment of the sample surface produces a mechanically destroyed surface layer. An increase in the incidence grazing angle leads to greater probing depth and hence participation of more perfect layers in forming the reflected radiation beam.

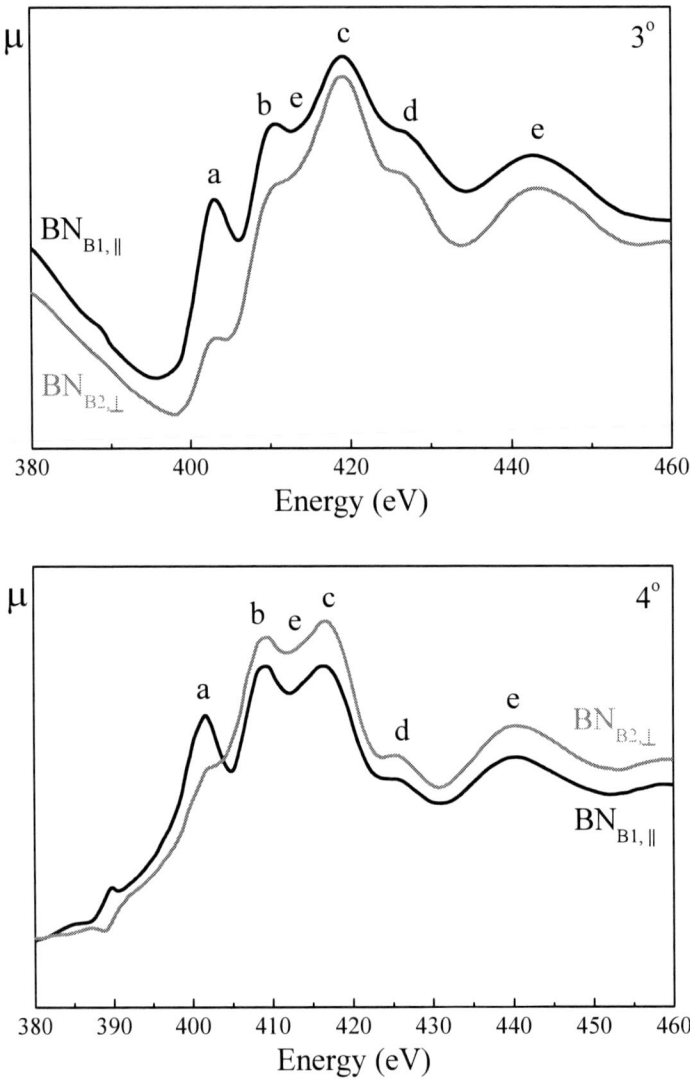

Figure 33. N K-absorption spectra of h-BN calculated from reflection spectra obtained for grazing incidence angles 3° and 4° for crystal orientations B_1 and B_2 using s-polarized radiation. The black curve denotes the data for the orientation B_1 ($BN_{B1,\parallel}$) and the grey curve denotes the data for orientation B_2 ($BN_{B2,\perp}$) (a) $\theta = 3°$ (b) 4° [110].

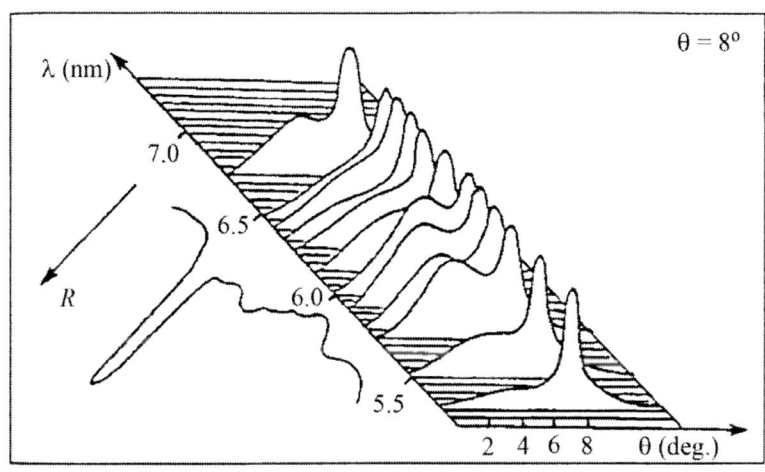

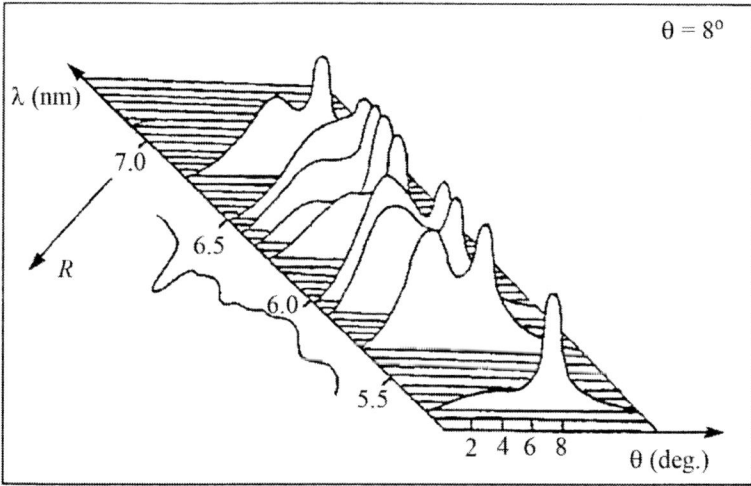

Figure 34 Scattering indicatricies from different orientations of the crystal h-BN [111]. Upper figure corresponds to the orientation A ($BN_{A,\perp,\parallel}$) and lower figure corresponds to orientation B_2 ($BN_{B2,\perp}$).

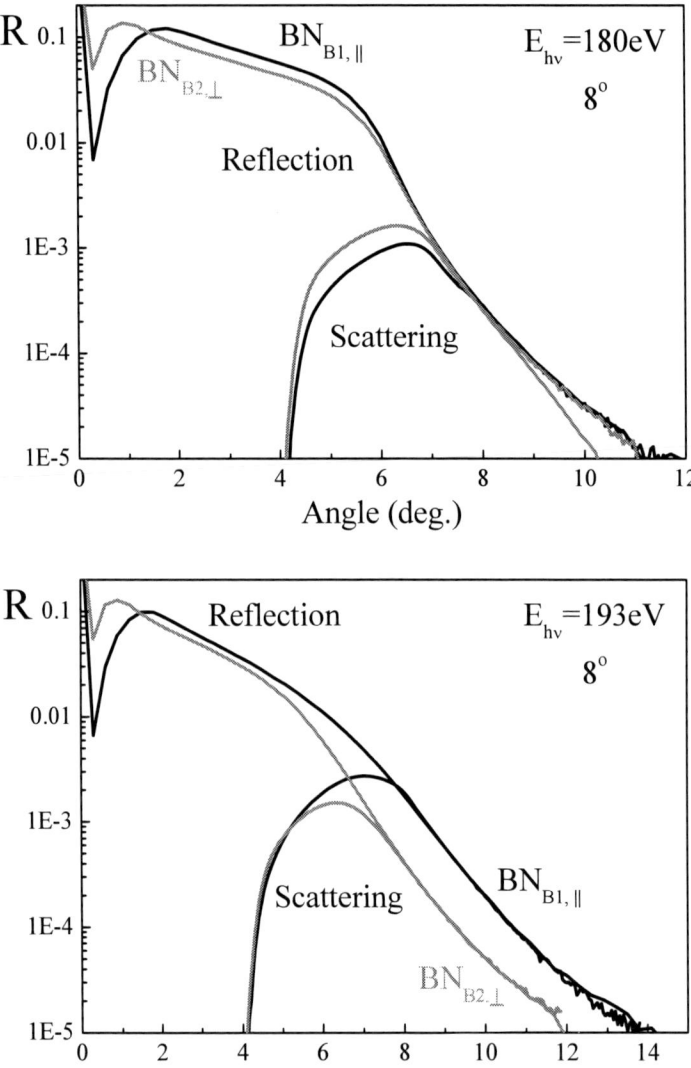

Figure 35. Angular dependences of the reflection coefficient and angular distribution of scattered radiation obtained for grazing incidence angle 8° for crystal orientations B_1 and B_2 using s-polarized radiation. The black curve denotes the data for orientation B_1 ($BN_{B1,\parallel}$) and the grey curve denotes the data for orientation B_2 ($BN_{B2,\perp}$). Energy: (a) E = 180 eV, (b) E = 193 eV [110].

Summarizing, it is necessary to point attention that strong intensity dependence due to crystal orientation exists for peak originated from the transitions B1s→ π states both in reflection and absorption spectra.

X-RAY REFLECTION FROM ROUGH SURFACES

To describe x-ray reflection from rough surfaces a number of theoretical approaches have been worked out [154-167]. The approach developed in [159,161,162] is based on the application of perturbation theory and on a rather general surface model. In approach [161,162] the boundary between substance and vacuum is described by the equation $z = \zeta(\mathbf{\rho})$, where ζ is a random function determining the statistical characteristics of the boundary. The plane $r = 0$ corresponds to an ideally smooth surface (i.e., average value $\langle \zeta(\mathbf{\rho}) \rangle = 0$) and the vector $\mathbf{\rho}$ lies in the (xy) plane.

The statistical properties of the surface are usually described by the correlation function of the surface heights $\chi(\mathbf{\rho}-\mathbf{\rho}') \equiv \langle \zeta(\mathbf{\rho})\zeta(\mathbf{\rho}') \rangle$ and the mean-square height $\chi(0) = \langle \zeta^2(\mathbf{\rho}) \rangle \equiv \sigma^2$. In the case of isotropic surfaces the correlation function of the surface heights depends only on $\chi(\rho) = \chi(|\mathbf{\rho}|)$. The characteristic scale of $\chi(\rho)$ variation is called the correlation radius a. The dielectric constant changes abruptly at the surface from 1 in vacuum to ε_+ in substance and can be expressed as:

$$\varepsilon(r) = 1 + (\varepsilon_+ - 1)H[z - \zeta(\vec{\rho})], \qquad (48)$$

Where $H(z)$ is step-like Heaviside function: $H(z > 0) = 1$ and $H(z < 0) = 0$. In general, the angular distribution of radiation scattered by surface (scattering diagram) is a function of two scattering angles $\Phi(\theta,\varphi) = (1/W_{inc})(dW_{scat}/d\Omega)$ and describes the power of radiation dW_{scat} scattered within a small solid angle (detector aperture) $d\Omega$ and normalized to the power of the incident beam W_{inc}. According to [8,9] the scattering diagram can be written as:

$$\Phi(\theta,\varphi) = \frac{2k^4}{\pi}\sin^3\theta_0 \, R_F(\theta_0)\frac{T(\theta)}{T(\theta_0)}\chi_F(\mathbf{q}-\mathbf{q}_0), \qquad (49)$$

where $R_F(\theta)$ and $T(\theta)$ make sense of Fresnel's reflection and trasmittion coefficients:

$$R_F(\theta) = \left|\frac{\sin\theta - \sqrt{\varepsilon_+ - \cos^2\theta}}{\sin\theta + \sqrt{\varepsilon_+ - \cos^2\theta}}\right|^2 ; \qquad (50)$$

$$T(\theta) = \left|\frac{2\sin\theta}{\sin\theta + \sqrt{\varepsilon_+ - \cos^2\theta}}\right|^2 ; \qquad (51)$$

$$\chi_F(\mathbf{q}-\mathbf{q}_0) \equiv \chi_B(\nu) = \int_0^\infty \rho\chi(\rho)J_0(\nu\rho)d\rho ; \qquad (52)$$

$$\nu = |\mathbf{q}-\mathbf{q}_0| ;$$

$J_0(\nu\rho)$ is Bessel function, θ_0 is grazing incident angle, θ and φ are grazing and azimuth scattering angles.

In typical laboratory scattering experiments the scattering diagram is integrated over the azimuth scattering angle and it is convenient to write

$$\Pi(\theta) = \frac{4k^3}{\sqrt{2\pi}}\sin^3\theta_0 \, R_F(\theta_0)\frac{T(\theta)}{T(\theta_0)}\chi_c(p); \qquad (53)$$

$$\chi_c(p) = \sqrt{\frac{2}{\pi}}\int_0^\infty \chi(\rho)\cos(p\rho)d\rho ; \qquad (54)$$

$$p = k|\cos\theta - \cos\theta_0|$$

Let's pay attention that the shape of the scattering diagram is defined by the product of two functions $T(\theta)$ and $\chi_c(p)$. $\chi_c(p)$ describes the surface relief and depends only on surface roughness statistics. $T(\theta)$ is defined only by optical properties of substance. It is clear that if functions $T(\theta)$ and $\chi_c(p)$ have any features they will be reflected in the shape of a scattering diagram. The function $\chi_c(p)$ has a maximum in the specular direction at $\theta=\theta_0$. At the same time the function $T(\theta)$ archives its maximum at the critical angle θ_c of total external reflection when the absorption is negligible and sharply decreases to asymptotical value for grazing angles $\theta>\theta_c$. That means that at special conditions the scattering diagram presents a peak in the angular distribution at angle located near the critical angle, limiting the region of total reflection. This peak is named Yoneda peak, which was discovered in 1963 year [168]. From theoretical point of view, according to [161-164], the experimental observation of the anomalous scattering peak is possible under certain conditions:

1. Existence of surface heights with small correlation radii:

$$a \leq \frac{\lambda}{\pi \theta_c^2} \quad (\lambda \text{ is radiation wavelength})$$

2. Realization of optimal glancing angle θ_0 value:

$$\frac{\delta \theta}{\theta_c} << \frac{(\theta_0 - \theta_c)}{\theta_0} \quad (\delta\theta \text{ is angle resolution})$$

3. Low absorption of radiation in substance:

$$\text{Im}\,\varepsilon_+ << \text{Re}(1-\varepsilon_+)$$

The last requirement leaves it open to doubt if observation of the Yoneda effect is possible in the ultrasoft x-ray region (typically absorption is strong in this region).

In [161-162] the model (1) was complicated by introduction of a surface transition layer. This introduction made the model more realistic and reflected the great differences of solids atomic and electronic structure in the interior volume and in the near-surface region. The dielectric constant and thickness of a surface transition layer strongly depend on the nature of the substance, surface preparation and treatment, external conditions, etc. It was shown that the existence of a surface transition layer does not affect the scattering of x-rays in the range $\theta<\theta_c$ but can significantly increase the intensity of the anomalous scattering peak in the case $\theta>\theta_c$.

So, from a theoretical point of view, the magnitude of the Yoneda peak depends basically on three characteristics of the sample: statistics of surface heights, x-ray absorption coefficient value, and features of surface transition layer. Yoneda peak angular position coincides with θ_c. and depends not only on the absorption coefficient but also on the wavelength of radiation. In the range of soft x-rays the existence of Yoneda effect was shown experimentally for the first time on hexagonal BN in our works [111,116].

X-RAY SCATTERING

Figure 34 shows the angular distribution of scattered radiation obtained for a grazing incidence angle $\theta_0 = 8°$ from two crystal faces (A and B_2(figure28)) as a function of wavelength, in the region of the K-absorption edge of boron. As can be seen, all indicatrices have the anomalous scattering peak (Yoneda peak) and its intensity depends on the wavelength λ, but the tendencies of changing are the same for each surface. The wavelength $\lambda = 6.49$ nm is the exception. This value corresponds to the position of the π - resonance of the B K-absorption spectrum (peak a , Fig. 32), which is strongly dependent on sample orientation. As was established above, the intensity of peak a increases sharply when the radiation reflects from crystal face A. According to the theoretical estimations, anomalous intensity is inversely related to the magnitude of absorption. This means that for a low value of absorption in peak a (realized in B_2 orientation) one can expect a large intensity of the Yoneda maximum and for a high magnitude of this peak (A) one can expect a low intensity of anomalous scattering. Such appropriateness is traced in experimental scattering distributions going from A orientation to B_2. It is possible to assert that the found out change of the intensity of anomalous peak is display by orientation dependence of crystal. The nature of this dependence is connected with orientation dependence of absorption coefficient and specificity of formation of anomalous scattering (Yoneda effect).

As follows from figure 29 B K-reflection spectra for different crystal orientations of the same surface at small incidence angles are differ significantly in absolute values of the reflection coefficients. The reflection coefficients for orientation B_1 are approximately 1.5-fold greater than for orientation B_2 at all energies. Figure 35 shows angular dependences of the reflection coefficient and angular distributions of scattered radiation for a grazing incidence angle $\theta = 8°$ from crystal orientations B_1 and B_2 obtained at the energies $E = 180$ eV (range of normal dispersion) and $E = 193$ eV (in the region of π resonance).

As can be seen the reflectivity for face B_1 is larger than for face B_2 in the region of normal dispersion *(E = 180 eV)*. The integral scattering intensities from both faces are similar in shape, but in the case of crystal orientation B_2 it is larger. Because the same surface was investigated in both cases it would appear reasonable that the revealed distinction connects with the planar anisotropy of the surface. A different situation arises with the energy $E = 193$ eV. In this case the reflectivity and integral scattering intensity both are larger for the crystal orientation B_1. Moreover the difference in anomalous scattering peak positions for the crystal orientations B_1 and B_2 should be noted. The difference measures 8°. Taking into account the connection of the angular position of the anomalous scattering peak with the critical angle of total external reflection (anomalous scattering peak angular position coincides with critical angle [116]) it is reasonably safe to suggest that the critical angles for two different crystal orientations of the same surface (B_1 and B_2) in the region of π resonance are distinguished on 1°.

Chapter 8

X-RAY POLAROID

Following the consideration presented above one can show that for anisotropic crystals there are two different critical angles for two different polarizations of the incident beam. In the case of the crystal surface is perpendicular to the optical axis the critical angle θ_s^{cr} for s-polarization is determined by $\varepsilon_\perp$ and θ_p^{cr} is determined by $\varepsilon_\parallel$ components of permittivity. Thus one can choose the glancing incidence angle to be less than θ_s^{cr} and more than θ_p^{cr} if $\theta_s^{cr} > \theta_p^{cr}$ or to be more than θ_s^{cr} and less than θ_p^{cr} if $\theta_s^{cr} < \theta_p^{cr}$:

$$\min\left(\theta_s^{cr}, \theta_p^{cr}\right) < \theta < \max\left(\theta_s^{cr}, \theta_p^{cr}\right) \tag{55}$$

In this case the s- (or p-) polarized radiation will decrease strongly in the sample than p- (or s-) radiation Because of small difference between the values of critical angles θ_s^{cr} and θ_p^{cr} the incident beam will be collimated. This effect can be used for creation of the x-ray Polaroid.

Mashavariani proposes in the work [117] the following scheme: x-ray radiation passes through the plane parallel sample provided that the angle between the incident wave and the crystal face is near to the critical value for the total external reflection. For the case where the glancing incidence angle is chosen according to (19), transmitted radiation will be strongly polarized.

In the case where the x- and y-axes belong to the crystal face and y- and z-axes belong to the scattering plane the formulae for transmission coefficients for s- (T_s) and p- polarized (T_p) radiation was received by Mashavariani in the work [117]:

$$T_s = \left| \frac{4\kappa_z^{(s)}(\omega/c)\sin\theta}{((\omega/c)\sin\theta + \kappa_z^{(s)})^2 \exp(-i\kappa_z^{(s)}L) - ((\omega/c)\sin\theta - \kappa_z^{(s)})^2 \exp(+i\kappa_z^{(s)}L)} \right|^2 \quad (56)$$

$$T_p = \left| \frac{4\varepsilon_{yy}\kappa_z^{(p)}(\omega/c)\sin\theta}{(\varepsilon_{yy}(\omega/c)\sin\theta + \kappa_z^{(p)})^2 \exp(-i\kappa_z^{(p)}L) - (\varepsilon_{yy}(\omega/c)\sin\theta - \kappa_z^{(p)})^2 \exp(+i\kappa_z^{(p)}L)} \right|^2 \quad (57)$$

$$\kappa_z^{(s)} = \frac{\omega}{c}\sqrt{\varepsilon_{xx} - \cos^2\theta} \quad (58)$$

$$\kappa_z^{(p)} = \frac{\omega}{c}\sqrt{\varepsilon_{yy} - \frac{\varepsilon_{yy}\cos^2\theta}{\varepsilon_{zz}}} \quad (59)$$

Let us now consider the interaction of different polarized radiation with the surface cut perpendicular to the optical **c**-axis of the h-BN crystal in the vicinity of BK-absorption edge. Figure 36 shows the B K-reflection spectra of h-BN measured for grazing incidence angles θ =4° and 20°. As follows from the figure the reflection spectra obtained with help of different polarization are quite the same in number of main spectral features and the energy positions of these structures are independent on the type of the polarization. For small grazing incidence angle θ =4°, the shape of the reflection spectra depends only slightly on the type of the polarization. Growth of the grazing incidence angle to 20° leads to a significant changing in the shape of all spectra. Strong intensity dependence due to type of the polarization for peaks A and B takes place.

The growth of the grazing incidence angle leads to considerable increasing of the intensity of peak A in the case of p-polarized radiation and the intensity of peak B in the case of s-polarized radiation. Remind that according to dipole selection rules with the regard to the crystal symmetry, in the simplest case of s-polarized radiation the absorption in h-BN must originate from transitions from B 1s to σ states in the case when $\mathbf{E}\perp\mathbf{c}$ and from B 1s to π states when

$\mathbf{E} \parallel \mathbf{c}$. According to experimental alignment the geometry $\mathbf{E} \perp \mathbf{c}$ was realized for s-polarized radiation and only transitions from B 1s states to σ states are expected in this case, so that the structure a is in agreement with our observations. On contrary when p-polarized radiation is used, the contribution of each component ($\mathbf{E} \parallel \mathbf{c}$ and $\mathbf{E} \perp \mathbf{c}$) to the overall reflection process is analyzed that leads to the appearance of intensive π-peak.

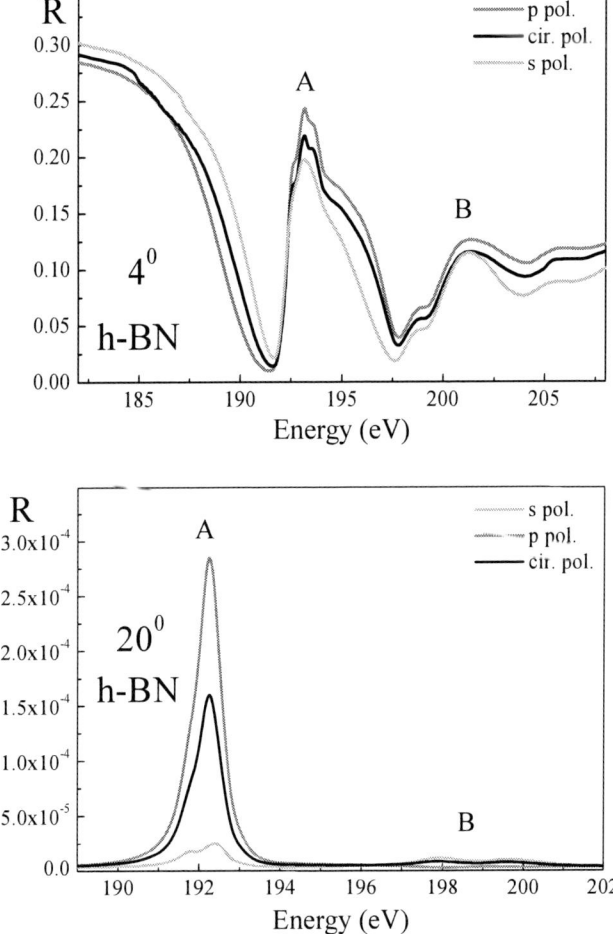

Figure 36. B K-reflection spectra of h-BN crystal obtained for $\theta = 4°$ and $20°$ with the help of s- and p-linearly polarized and circular (positive) polarized radiation [118].

In the figure 37 the angular and energy dependences of the coefficient $P = (R_s - R_p)/(R_s + R_p)$ are plotted. One can see from the figure that for the energy region from 185 eV to 198 eV there is an untraditional strong polarization effect. For small grazing incidence angle $\theta = 4°$ this energy region falls to two parts. Inside the first one the R_s dominates considerably. Within the second one p-polarized character of the reflectivity is expressed strongly. The growth of the angle θ intensifies p-polarized dependencies of the reflection coefficient. Analysis of the measured angular dependencies of the reflectivity shows that the critical angle for s-polarized radiation is smaller in comparison with p-polarization and difference between values of critical angles makes approximately 0.8°.

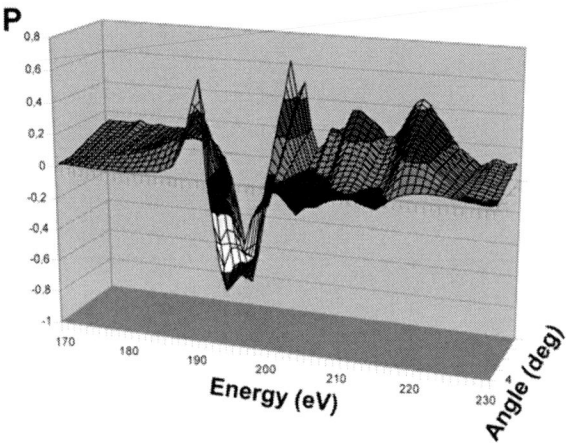

Figure 37. Angular and energy dependences of $P = (R_s - R_p)/(R_s + R_p)$ [118].

So, follows the scheme proposed by Mashavariani [117] one can calculated the polarization coefficient $P = (T_s - T_p)/(T_s + T_p)$. This coefficient shows the degree of polarization coefficient of transmitted radiation if the incident radiation is completely unpolarized. Thus, if $P = +1$ then the transmitted polarization has s polarization, if $P = -1$, then the transmitted polarization has p polarization, and if $P = 0$ the transmitted polarization is completely unpolarized. Such calculations were carried out by Mashavariani in the work [117] for different glancing incidence angles and different thicknesses of the sample in the case when the optical axis of the crystal is perpendicular to the crystal surface. Figure 38 the results of calculations are presented.

One can see from the figure that in the vicinity of BK-absorption edge there is a strong polarization effect, and the transmitted radiation will have s polarization (P = +1). One can also see from the figure that in the case of small glancing incidence angles the values of P are closer to +1 or to -1, and thereby small glancing incidence angles are more convenient for the creation of x-ray polaroids.

As follows from the figure 39 even for small thickness of sample there is a strong x-ray Polaroid effect and x-ray Polaroid can be made even using thin films. Also, from the figure 39 follows that the intensity of the transmitted radiation is quite sufficient for practical use.

In conclusion, the electronic and atomic structure of h-BN films and X-ray interactions with them are not considered and discussed in the work. In the same time we stress that X-ray investigations of the films (see, e.g. [119-121]) are very important for deeper understanding of electro-optical properties of the h-BN crystal near the B and N K edges.

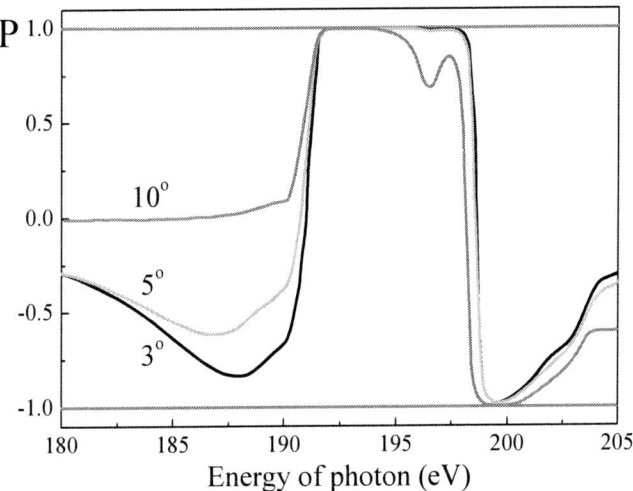

Figure 38. The polarization coefficient P for h-BN crystal for the glancing angles $\theta=3°$, $5°$ and $10°$ versus the photon energy. The optical axis is perpendicular to the crystal face. The thickness of the sample is L D 0.03 µm.

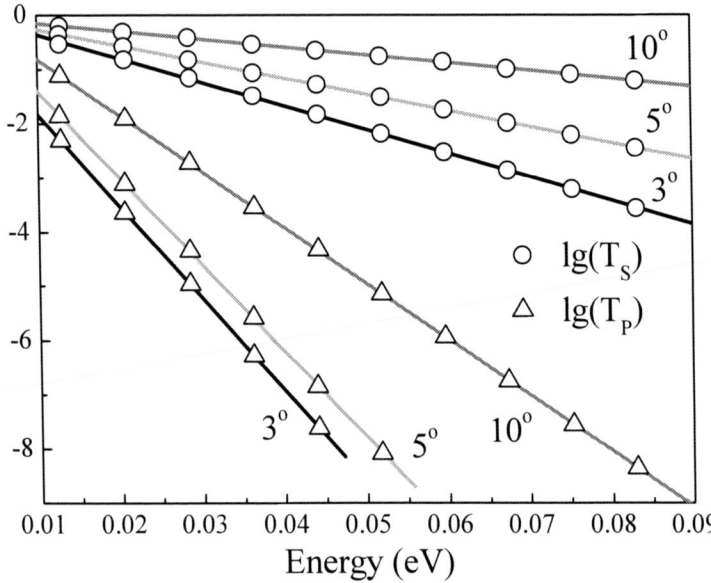

Figure 39. lg(T_s) / and lg(T_p) for θ=3°, 5° and 10° versus the thickness of the sample for a photon energy of 194 eV.

REFERENCES

[1] Edgar, J. H. *Properties of Group III Nitrides* 1994, London: INSPEC.
[2] Ariyama, K.; Mase, S. *Busseiron Kenkyu* 1950, 33, 109.
[3] Ariyama, K.; Mase, S. *Progr. Theoret. Phys. (Kyoto)* 1954, 12, 244; 1954, 12, 246.
[4] Larach, S.; Shrader, R. E. *Phys. Rev. B* 1956, 104, 68.
[5] Halpren, V. *J. Phys. Chem. Solids* 1963, 24, 1495.
[6] Fomichev, V. A. *Izv. Akad. Nauk SSSR*, Ser. Fiz. 1967, 31 (6), 957; *Fiz. Tverd. Tela* (Leningrad) 1967, 9 (11), 3167 [Sov. Phys. Solid State 1967, 9, 2496]; *Fiz.Tverd. Tela* (Leningrad) 1971, 13, 907) *Sov. Phys. Solid State* 1971, 13, 754]; Fomichev, V. A. and Rumsh, M.A., *J.Phys.Chem.Solids* 1968, 29, 1015.
[7] Rand, M. J.; Roberts, J. F. *J. Electrochem. Soc.* 1968, 115, 423.
[8] Doni, E.; Parravicini, G. P. *Nuovo Cimento B* 1969, 64, 117.
[9] Savostina, S. N.; Tulvinsky, V. B.; Guzhov, A. A. *Proceedings of VII Ural Conference on Spectroscopy* [IFM UHTs AN SSSR, Sverdlovsk, USSR] 1971, 189±191 (in Russian).
[10] Nakhmanson, M. S.; Smirnov, V. P. *Fiz. Tverd. Tela* (Leningrad) 1971, 13, 905 [Sov. Phys. Solid State 1971, 13, 752]; *Fiz. Tverd. Tela* (Leningrad) 1971, 13, 3288 [Sov. *Phys. Solid State* 1971, 13, 2763]; Sov. *Phys. Solid State* 1972, 13, 2763.
[11] Zupan, J., *Phys.Rev. B* 1972, 6 (6), 2477.
[12] Khusidman, M. B., *Fizika Tverdogo Tela* [Sov. *J. Solid State Phys.*] 1972, 14, 3287.
[13] Zupan, J.; Kolar, D. *J. Phys. Chem.* 1972, 5, 3097.
[14] Baronian, W., *Mater. Res. Bull.* 1972, 7, 119.

[15] Katzir, A.; Zunger, A.; Halperin, A., *Bull. Amer. Phys. Soc.* 1976, 21, 246.
[16] Leapman, R. D.; *Silcox Phys.Rev.Lett.* 1979, 42, 1361; Leapman, R. D.; Fejes, P. L., Silcox J. *Phys. Rev. B* 1983, 28, 2361.
[17] Dovesi, R.; Pisani, C.; Roetti, C. *Int. J. Quantum Chem.* 1980, 17, 517.
[18] Carpenter, L. G.; Kirby P. Y. *J. Phys. D* 1982, 15, 1143.
[19] Robertson, J. *Phys.Rev.B* 1984 29(4), 2131.
[20] Hoffman, D. M.; Doll, G. L.; Eklund, P. C. *Phys. Rev. B* 1984, 30, 6051.
[21] Xu, Y. N.; Ching, W. Y. *Phys. Rev. B* 1991, 44 (15), 7787.
[22] Lukomskii, A. J.; Shipilo, V. B.; Gameza, L. M. *J. Appl. Spectrosc.* 1993, 57, 607.
[23] Grinyaev, S. N.; Lopatin, V. V. *Zh. Strukt. Khim.* 1997, 38 (1), 32.
[24] Abdellaoui, A.; Bath, A.; Bouchikhi, B.; Baehr, O. *Mater. Sci. Eng.* B 1997, 47, 257.
[25] Loh, K. P.; Sakaguchi, I.; Gamo, M. N.; Tagawa, S.; Sugino, T.; Ando, T. *Appl. Phys. Lett.* 1999, 74, 28.
[26] Solozhenko, V. L.; Lazarenko, A. G.; Petitet, J. P.; Kanaev, A. V. *J. Phys. Chem. Solids* 2001, 62, 1331.
[27] Ilyasov, V. V.; Zhdanova, T. P.; Nikiforov, I. Ya. *Physics of the Solid State* 2003, 45, No. 5, 816. [*Fiz. Tverd. Tela* 2003 45, No. 5, 778].
[28] Watanabe, K.; Taniguchi, T.; Kanda, H. *Nat. Mater.* 2004, 3, 404.
[29] Ooi, N.; Rairkar, A.; Lindsley, L.; Adams, J. B. J. Phys.: Condens. Matter 2006, 18, 97.
[30] Lelonis, D. A. et al 2003 *General Electric Company Publication* No 81506.
[31] *Handbook of Chemistry and Physics* 2005 (Boca Raton, FL: CRC Press).
[32] Lelonis, D. A. *General Electric Company Publication* 2003 No 81505.
[33] Meunier, V.; Roland, C.; Bernholc, J.; Nardelli, M. B. *Appl. Phys. Lett.* 2002, 81, 46.
[34] Will, G.; Perkins, P.G. *Materials Letters* 1999, 40(1), 1.
[35] Watanabe, S.; Miyake, S.; Murakawa, M. *Surf. Coat. Technol.* 1991, 49, 406.
[36] Kimura, Y.; Wakabayashi, T.; Okada, K.; Wada, T.; Nishikawa, H. *Wear* 1999, 232, 199.
[37] Lipp, A.; Schwetz, K. A.; Hunold, K. *J. Eur. Ceram. Soc.* 1989, 5, 3.
[38] Zhang, Y.; He X.; Han, J.; Du, S. *J. Mater. Process. Technol.* 2001, 116, 161.
[39] Choi, B. J. *Mater. Res. Bull.* 1999, 34, 2215.

[40] Yamamura, S.; Takata M.; Sakata, M. *J. Phys. Chem. Solids* 1997, 58, 177.
[41] Muramatsu, Y.; Kaneyoshi, T.; Gullikson, E. M.; Perera, R. C. C. *Spectrochim. Acta* A 2003, 59, 1951.
[42] Hamrim, K, Johanson, G.; Gelius, U.; Nordling, C.; Siegbahn, K. *Phys. Scripta.* 1970, 1, 277.
[43] Hendrickson, D. M.; Hollander, J. M., *Inorg. Chem.* 1969, 8, 2642.
[44] Franke, R.; Bender, St.; Hormes, J.; Fresenius, J. *Anal. Chem.* 1996, 345, 874.
[45] Zunger, A.; Katzir, A.; Halperin, A. *Phys.Rev.* 1976, B13, 5560.
[46] Zunger, A. *J.Phys.* 1974, C7, 76 and 96.
[47] Funakawa, S.; Yamamuro, Yu.; Luo, H.; Sugino, T. *Diamond and Related Materials*, 2004, 13(4-8), 994-8.
[48] Busta, H. H.; Pryor, R. W. *J. Vac. Sci. Technol. B* 1998, **16,** 1207.
[49] *Ceramic Industry Materials Handbook*, January 1998, 79.
[50] Lelonis, D. A.; Tereshko, J. W.; Andersen, C. M. *Momentive Performance Materials Inc.*, 2006-2007, www.momentive.com.
[51] Sugino, T.; Tanioka, K.; Kawasaki, S.; Shirafuji, J. Diamond *Relat. Mater.* 1998, 7, 632.
[52] Tarrio, C.; Schnatterly, S. E. *Phys. Rev. B* 1989, 40, 7852.
[53] Muramatsu, Y.; Kawai, J.; Scimeca, T.; Oshima, M.; Kato, H. *Physica Scripta* 1994, 50, 25.
[54] Pavlychev, A. A.; Vinogradov, A. S.; Stepanov, A. P.; Shulakov, A. S. *Opt. Spectrosc.* 1993, 75, 554.
[55] Pavlychev, A. A.; Vinogradov, A. S.; Akimov, V. N.; Nekipelov, S. V., *Physica Scripta*, 41 (1990) 160.
[56] Pavlychev, A. A.; Rühl, E., J. Electron. Spectrosc. *Relat. Phenom.*, 2000 106, 207; ibid 2000, 107, 203.
[57] Vinogradov, A. S.; Nekipelov, S. V.; Pavlychev, A. A., Sov. *Solid State Phys.* 1991, 33, 508.
[58] Franke, R.; Bender, S.; Hormes, J.; Pavlychev, A. A.; Fominykh, N. G., *Chem. Phys.*, 1997, 216, 243.
[59] Pavlychev, A. A.; Franke, R.; Bender, St.; Hormes, J., *J. Phys. Condens. Matter* 1998, 10, 2181.
[60] Pavlychev, A. A.; Barry, A.; Vinogradov, A.S., *Sov. Solid State Phys.* 33 (1991) 2985.
[61] Pavlychev, A. A.; Hallmeier, K.-H.; Hennig, C.; Hennig, L.; Szargan, R., *Chem. Phys.* 1995, 201, 547.

[62] Heine, V.; et.al. in Solid State Physics. *Advances in Research and Applications*. Eds. H. Ehrenreich, F. Seitz, D. Turnbull. Academic Press. NY – London, 1970, 24.
[63] Adams, W. H. *J. Chem. Phys.* 1962, 37, 2009.
[64] Gilbert, T. L. *Molecular orbitals in chemistry, physics and biology*, (Academic Press, NY, 1964) p. 405.
[65] Babikov, V. V. *Phase functions method in quantum mechanics*. (Nauka, Moscow, 1976).
[66] Calogero, F. *The variable phase approach to potential scattering* (Academic Press, NY, 1972).
[67] Pavlychev, A. A.; Fominykh, N. G. *J. Phys. Condens. Matter* 1996, 8, 2305.
[68] Abarenkov, I. V.; Heine V. *Phil. Mag.* 1965, 12, 529.
[69] Demkov, Yu. N.; Ostrovsky, V. N. *Zero-range potential approximation in atomic physics* (LGU, Leningrad, 1974).
[70] Amusia, M. Ya.; Baltenkov, A. S.; Dolmatov, V. K.; Manson, S. T.; Msezane A. Z. *Phys. Rev. A* 2004, 70, 023201.
[71] Piancastelli, M. N. *J. El. Spectrosc. Relat. Phenom.* 1999, 100, 167.
[72] Pavlychev, A. A.; Brykalova, X. O.; Flesch, R.; Rühl, E. *Phys. Chem. Chem. Phys.* 2006, 8,1914.
[73] Pavlychev, A. A.; Brykalova, X. O.; Mistrov, D. A.; Flesch, R., Rühl, E. *J El Spectrosc. Relat. Phenom.* 2008, 166-167, 45.
[74] Pavlychev, A. A.; Flesch, R.; Rühl, E. *Phys. Rev. A* 2004, 70, 015201.
[75] Chaiken, A.; Terminello, L. J.; Wong, J.; Doll, G. L.; Tayllor, C. A. *Appl. Phys. Lett.* 1993, 63, 2112.
[76] Terminello, L. J.; Chaiken, A.; Lapiano-Smith, D. A.; Doll, G. L.; Saito, T. *J. Vac. Sci. Technol.* A 1994, 12, 2462.
[77] De Fanis, A.; Saito, N.; Pavlychev, A. A., et. al. *Phys. Rev. Lett.* 2002, 89, 023006.
[78] Mistrov, D. A.; De Fanis, A.; Kitajima, M.; Hoshino, M.; Shindo, H.; Tanaka, T.; Tamenori, Y.; Tanaka, H.; Pavlychev, A. A.; Ueda, K. *Phys. Rev. A* 2003, 68, 022508.
[79] Pavlychev, A. A.; Mistrov, D. A. *J. Phys. B: At. Mol. Opt. Phys.* 2009, 42, 055103.
[80] Brykalova, X. O.; Flesch, R.; Seadarogu, E.; Blobner, F.; Feulner, P.; Pavlychev, A.A.; Rühl, E. Book of Abstracts, 37-th Int. Conf. on Vaccum Ultraviolet and X-ray Physics, Vancouver, Canada, 11-16 July, 2010.

[81] Ishiguro, E.; Iwata, S.; Suzuki, Y.; Mikuni, A.; Sasaki, T. *J. Phys. B: At. Mol. Phys.* 1981, 14, 1841.
[82] Ueda, K.; De Fanis, A; Saito, A., et.al. *Chem. Phys.* 289 (2003) 135.
[83] Simon, M.; Ueda, K.; Lablanquie, P.; Lavolee, M.; Morin, P.; Kosugi, N., *Chem. Phys. Lett.* 1995, 238, 42.
[84] Ueda, K.; Ohmori, K.; Okunishi, M.; Chiba, H.; Shimizu, Y.; Sato, Y.; Hayaishi, T.; Shigemasa, E.; Yagishita, A., *Phys. Rev. A* 1995, 52, R1815.
[85] Tanaka, S.; Kayanuma, Y.; Ueda, K., *Phys. Rev. A* 1998, 57, 3437.
[86] Flesch, R.; Pavlychev, A. A.; Neville, J.J.; Blumberg, J.; Kuhlmann, M.; Tappe, W.; Senf, F.; Schwarzkopf, O.; Hitchcock, A. P.; Rühl, E. *Phys. Rev. Lett.* 2001, 86, 3767.
[87] Ma, Y.; Skytt, P.; Wassdahl, N.; Glan, P.; Mancini, D. C.; Guo, J.; Nordgren, J., *Phys. Rev. Lett.* 1993, 71, 3725.
[88] Mansour, A.; Schnatterly, S. E. *Phys. Rev. Lett.* 1987, 59, 567.
[89] O'Brien, W. L.; Dong, J.; Jia, Q-Y, et.al. *Phys. Rev. Lett.* 1993, 70, 238.
[90] Eberhard, W. *Phys. Scripta* 1987, 17, 28.
[91] Morin, P.; Nenner, I. *Scripta* 1987, 17, 171.
[92] Ueda, K.; Chiba, H.; Sato, Y.; Hayaishi, T.; Shigemasa, E.; Yagishita A. *Phys. Rev. A* 1992, 46, R5.
[93] Filatova, E.; Pavlychev, A. A.; Blessing, C.; Friedrich, J. *Physica B* 1995, 208+209, 417.
[94] Wada, T.; Yamashita, N. J. *Vac. Sci. Technol. A* 1992, 10, 515.
[95] Jimenez, I.; Jankowski, A.; Terminello, L. J.; Carlisle, J. A.; Sutherland, D. G.; Doll, G. L.; Mantese, J. V.; Tong, W. M.; Shuh, D. K.; Himpsel, F. *J. Appl. Phys. Lett.* 1996, 68, 2816.
[96] Catellani, A.; Pasternack, M.; Balderechi, A.; Freeman, A. J., *Phys. Rev. B* 1987, 36, 6105; Catellani, A.; Pasternack, M.; Balderechi, A.; Jansen, H. J. F.; Freeman, A. *J. Phys. Rev. B* 1985, 32, 6997.
[97] Park, K. T.; Terakura, K.; Hamada, N. *J. Phys. C: Solid State Phys.* 1987, 20, 1241.
[98] Chadi, D. J.; Chang, K. J. *Phys. Rev B* 1989, 39, 10063.
[99] Suzuki, S.; Bower, C.; Kiyokura, T.; Nath, K. G.; Watanabe, Y.; Zhou, O. J. Electron *Spectrosc. Relat. Phenom.* 2001, 114, 225.
[100] Kramberg, C.; Rauf, H.; Shiozawa, H.; Knupfer, M.; Buchner, B.; Pichler, T.; Batchelor, D.; Kataura, H. *Phys. Rev. B* 2007, 75, 235437.
[101] Nekipelov, S. V.; Akimov, V. N.; Vinogradov, A.S. *Sov. Solid State Phys* 1991, 33, 663.

[102] Nekipelov, S. V.; Akimov, V. N.; Vinogradov, A.S., *Sov. Solid State Phys* 1988, 30, 3647.
[103] Born, M.; Wolf, E. *Principles of Optics* 2002, Cambridge.
[104] Sirotin, Yu. I.; Shaskolskaya, M. P. *Fundamentals of crystal physics* 1979, M., Nauka.
[105] Tegeler, E.; Kosuch, N.; Wiech, G.; Faessler, A. *Phys. Stat. Sol. (b),* 1979, 91(1) 223.
[106] Terauchi, M.; *Ultramicroscopy* 2006, 106, 1069.
[107] Barth, J.; *et al DESY-SR* 1980, 11, 10.
[108] Kawaguchi, M.; Kuroda, S.; Muramatsu, Y. *J. of Phys. and Chem. of Solids* 2008, 69, 1178.
[109] Vedrinskii, R. V.; Kraizman, V. L.; Novakovich, A. A.; Machavariani, V. Sh. *J. Phys. Condens. Matter* 1993, 5, 8643.
[110] Filatova, E. O.; Lukyanov, V. A. *J. Phys.: Condens. Matter* 2002, 14, 11643.
[111] Filatova, E. O.; Blagoveshchenskaya, T. A. *J. X-Ray Scienc. & Technology*, 1993, 4 1.
[112] Filatova, E. O.; Stepanov, A. P.; Blessing, C.; Friedrich, J.; Barchewitz, R.; Andre, J.-M.; LeGuern, F.; Bac, S.; Troussel D. *J. Phys.: Condens. Matter*, 1995, 7 2731.
[113] Filatova, E. O.; Pavlychev, A. A.; Blessing, C.; Friedrich, J. 1995 *Physica* B 208/209 417.
[114] Filatova, E.; Lukyanov, V.; Barchewitz. R.; Andre, J.-M.; Idir. M.; Stemmler. Ph. *J. Phys.: Condens. Matter*, 1999, 11 3355.
[115] Filatova E. O.; Shulakov A. S. *J. Colloid Interface*, 1995, Sci. 169361.
[116] Filatova E. O.; Blagoveshenskaya. T. A., *J. X-Ray Sci. Technol.*, 1992, 3, 204.
[117] Machavariani, V. Sh. *J. Phys.: Condens. Matter* 1996, 8, 10687.
[118] Filatova, E. O.; Taracheva, E. Yu.; Andr´e, J.-M.; Mertins, H.-Ch.; Abramsohn, D. *J. of Electron Spec. and Rel. Phenomena* 2005, 144–147, 937.
[119] Preobrajenski, A. B.; Vinogradov, A. S.; Martensson, N. *Phys. Rev. B.* 2004, 70, 165404.
[120] Auswarter, W.; Kreutz, T. J.; Greber, T.; Osterwalder, J. *Surf. Sci.* 1999, 429, 226.
[121] Grad, G. B.; Blaha, P.; Auswarter, W.; Greber, T. *Phys. Rev. B* 2003, 68 035404.
[122] Sheehy, J. A.; Gill, T.J.; Winstead, C.L.; Farren, R. E.; Langhoff, P.W. *J. Chem. Phys.* 1989, 81 3647; 1987, 86 3253.

[123] Bayliss, N.S.; *J. Chem. Phys.* 1948, 16 287.
[124] Kuhn, H; *Helv. Chim. Acta*, 1948, 31 1441.
[125] Pavlychev, A.A.; Brykalova X.O.; R. Flesch,; E. Rühl, to be published.
[126] Pavlychev, A. A.; Barry, A.; Vinogradov, A. S., *Physica Scripta*, 1991, 44 399
[127] Pavlychev, A.A.; Fominych, N.G.; Watanabe, N.; Soejima, K.; Shigemasa, E.; Yagishita, A., *Phys. Rev. Lett.*, 1998, 81 3623
[128] Pavlychev, A. A.; R\uhl, E.; *J. Electron. Spectrosc. Relat. Phenom.*, 2000, 106 207
[129] Ueda, K.; Science, 2008, 320 884
[130] Schöffler, M. S.; Titze, J.; Petridis, N.; Jahnke, T.; Cole, K.; Schmidt, L. Ph. H.; Czasch, A.; Akoury, D.; Jagutzki, O.; Williams, J. B.; Cherepkov, N. A.; Semenov, S. K. McCurdy, C. W.; Rescigno, T. N.; Cocke, C. L.; Osipov, T.; Lee, S.; Prior, M. H.; Belkacem, A. ; Landers, A. L.; Schmidt-Böcking, H.; Weber, Th.; Dörner, R. *Science*, 2008, 320 920
[131] Cederbaum, L.S., *Annals of Physics*, 2001, 291 169
[132] Wu, T.U.; Ohmura, N.; *Quantum Theory of scattering*, 1962, Prentice-Hall Inc, NY.
[133] Pavlychev, A. A.; *J. Phys. B: At. Mol. Opt. Phys.* 1999, 32 2077
[134] Ueda, K.; Hoshino, M.; Kitajima, M.; Tanaka, H.; De Fanis, A.; Saito, N.; Pavlychev, A. A.; AIP Conference Proceedings, 2003, #697, p.158, NY, Melville
[135] Hoshino, M.; Tanaka, T.; Kitajima, M.; Tanaka, H.; De Fanis, A.; Pavlychev, A. A.; Ueda, K. *J. Phys. B: At. Mol. Opt. Phys.* 2003, 36 L381.
[136] Hitchcock, A.P.; Stör, J., *J. Chem. Phys.* 1987, 87 3253.
[137] Stör, J., *NEXAFS Spectroscopy*, Springer, Berlin, 1992
[138] Piancastelli, M. N.; Lindle, D. W., Ferrett, T. A., Shirley, D. A., *J. Chem. Phys.* 1987, 86 2765
[139] Fikhtengolz, G.M.; The course of differential and integral calculus. 1970, Vol.2. Nauka, M., (in russ.)
[140] Vedrinskii, R. V.; Kraizman, V.L.; Novakovich, A.A.; Teterin, Yu. A., preprint IAE-3368/12, Kurchatov Institute, M., 1980.
[141] Kraizman, V.L.; Vedrinskii, R. V., *Zh. Exp Theor. Phys.* (USSR) 1978, 74, 1215
[142] Doering, J. P.; Gedanken, A.; Hitchcock, A.P.; Fisher, P.; Moore, J.; Olthoff, J. K.; Tossell, J.; Raghavachari, K.; Robin, M.B. *J. Amer. Chem. Soc.* 1986, 108, 3602.

[143] Brzhezinskaia, M. M.; Vinogradov, A. S.; Krestinin, A. V.; Zvereva, G. I.; Kharitonov, F. P.; Kulakova G. I. *Fizika Tverd. Tela (Sol. St. Phys.)* 2010, 52, 819
[144] Toll, J. S. *Phys. Rev.* 1956, 104, 1760.
[145] Hagemann, H J; Clucher, R.; Nielson, U. *Prepint DESY* 41-73/10, 1973, 1.
[146] Stern, F. *Solid State Physics.* 1963, 15, ed. F Seitz and D Turnbull (London: Academic)
[147] Young, R H. *J. Opt. Soc. Am.* 1977, 67, 520
[148] Plaskett, J. *J. Chem. Phys.* 1963, 38, 612
[149] Vinogradov, A. V.; Zorev, N. N.; Kozhevnikov, I. V.; I.Sagitov, S.; Turyanskii, A. G. *Sov. Phys. JETP* . 1988, 67, 1631.
[150] Sheik-Bahae, M."Nonlinear optics basics. Kramers–Kronig relations in nonlinear optics," in Encyclopedia of Modern Optics, ed. R. D. Guenther, 2005, Academic.
[151] Lucarini, V.; Saarinen, J. J; Peiponen, K.-E.; Vartiainen, E.M. Kramers-Kronig Relations in Optical Materials Research, 2005, Springer.
[152] Filatova, Elena; Sokolov, Andrey; André, Jean-Michel; Schaefers, Franz; Braun, Walter. *Appl. Optics.* 2010, 49(14), 2539.
[153] Filatova E. O.; Vinogradov A. S.; Simkina T. M.; Sorokin I. A. *Fiz. Tverd.Tela* (Leningrad) 1985, 27(4), 991.
[154] Beckmann, P.; Spizzichino, A. *The Scattering of Electromagnetic Waves from Rough Surfaces* 1963 Pergamon, New York.
[155] Bennett, H. E.; Porteus, J. O. *J. Opt. Soc. Am.* 1961, 51, 123.
[156] Porteus, J. O. *J. Opt. Soc. Am.* 1963, 53, 1394.
[157] Hogrefe, H.; Kunz, C. *Appl. Opt.* 1987, 26, 2851.
[158] Elson, J. M. *Phys. Rev.* 1984, B30, 5460.
[159] Maradudin, A. A.; Mils D.L. *Phys.Rev.B.* 1975, 11, N4, 1392.
[160] Sinha, S. K.; Sirota, E. B.; Garoff, S.; Stanley, H. B. *Phys. Rev. B.* 1986, 38, №4, 2297.
[161] Vinogradov, A.V.; Zorev, N. N.; Kozhevnikov, I. V.; Sagitov, S. I.; Turyansky, A. G.. *Sov.Phys.JETP.* 1988, 67, 389.
[162] Vinogradov, A. V.; Zorev, N. N.; Kozhevnikov, I. V.; Yakushkin, I. G. *Zh. Eksp. Tear. Fiz.* 1985, 89(6), 2124.
[163] Andreev, A. V. *Sov. Phys. Usp.* 1985, 28, 70.
[164] *Andreev, A.V. Usp. Fiz. Nauk.* 1985, 145(1), 113.
[165] *Smirnov, L. A.*; *Sotnikova,* S.; Anokhin, B. S.; Taibin, B. Z. *Opt. Spectrosc.* (USSR), 1979, 46, 329

[166] *Smirnov, L. A.*; *Sotnikova,* S.; Kogan, YU. I. *Opt. spektrosk.* 1985, 58(2), 400.
[167] Rovinsii, B. M.; Sinaiskii, V. M.; Sidenko, V. I. *Sov. Phys. Solid Stale.* 1972, 14, 237.
[168] Yoneda Y. *Phys.Rev.* 1963, 131(5), 2010.
[169] Kronig de L.R., *Z. Physik.* 1931, 70, 317. Kronig de L.R., *Z. Physik.* 1932, 75, 468.
[170] Nefedov V.I., *J. Struct. Chem.* 1979, 11, 277.
[171] Gianturco F.A.; Guidotti M.; Lamanna U., *J. Chem. Phys.* 1972, 57, 840
[172] Dehmer, J.L., *J. Chem. Phys.* 1972, 56, 4496.

INDEX

A

absorption, vii, 1, 5, 6, 7, 9, 14, 15, 17, 18, 19, 20, 21, 22, 23, 25, 28, 32, 37, 38, 39, 40, 42, 43, 44, 46, 47, 48, 59, 63, 64, 69, 73, 74, 76, 77, 78, 81, 83, 84, 88, 91
absorption spectra, 14, 20, 21, 22, 38, 39, 40, 48, 59, 63, 73, 74, 76, 77, 78, 81
acid, 48
advantages, 49
amplitude, 27, 68
anisotropy, vii, 3, 40, 41, 85
asymmetry, 46
atomic layers, vii, 17
atomic orbitals, 10
atoms, vii, 1, 3, 4, 8, 9, 10, 11, 12, 13, 15, 17, 18, 19, 20, 21, 24, 28, 29, 30, 31, 32, 37, 40, 41, 43, 44, 45, 47, 48, 62, 75, 76, 77

B

B_2O_3, 38, 40, 41, 44
$B_3N_3H_6$, 38, 40, 41, 48, 75
band gap, 2, 9, 10, 14, 15
BBr_3, 18, 41, 43, 75
BCl_3, 18, 41, 43, 75
beams, 62
BF_3, 18, 38, 40, 41, 43, 44, 45, 75
binding energies, 7, 8, 11
binding energy, 8
B-N bonds, 2
bonds, 2, 3, 18, 48
boric acid, 2

C

calculus, 99
carbon, 1, 2, 46, 48
chemical bonds, 49
chemical properties, 2
chemical reactions, 2
chemical vapour deposition, 2
clusters, 1, 29, 32, 44
CO, 18, 20, 21
CO2, 18, 28, 29
coatings, 2
combustion, 2
competition, 30
composition, 15
compounds, vii, 2, 3, 9, 17, 18, 20, 21, 28, 38
conduction, 5, 9, 15, 46, 63
conductivity, 14
conductor, 2
configuration, 45, 76
configurations, 3, 35, 45, 76
contradiction, 30
coordination, 27, 48, 62, 77
core-hole localization, 9, 25, 36, 37, 38, 51

correlation, 14, 19, 21, 32, 42, 48, 81, 83
critical value, 87
crystal structure, vii
crystalline, 15, 29, 77
crystals, 41, 49, 55, 56, 57, 75, 87
cubic system, 18

D

damping, 32
decay, 29, 30, 44
decomposition, 42, 43
defects, 44
deformation, 44, 45, 76
degenerate, 76
delocalization, 29, 30
density functional theory, 11
detection, 23
deviation, 32, 34
DFT, 11, 14, 15
dielectric constant, 51, 52, 68, 81, 83
dielectric permittivity, 55
differential equations, 23
dispersion, 1, 69, 70, 85
displacement, 45, 51, 53, 76
dissociation, 43, 44
distortions, 33, 46, 76
dominance, 46

E

effective potential, 17, 18, 19
electric field, 51, 55, 61, 62, 64, 74
electromagnetic, 51, 54, 63
electromagnetic waves, 51, 63
electron, 5, 10, 14, 22, 23, 24, 26, 27, 29, 30, 31, 32, 33, 38, 40, 44, 62, 63, 75, 76
electronic structure, 5, 77, 83
electro-optical properties, 91
emission, 1, 5, 6, 7, 8, 9, 12, 13, 14, 29, 31, 44, 46, 56, 59
energy splitting, 18, 20, 41, 42, 43, 76
engineering, 2
environment effect, 18

equality, 31
equilibrium, 17, 20, 21, 41
equivalent atoms, 1, 28, 29, 30
excitation, 28, 30, 31, 32, 37, 43, 44, 45, 46, 76
extraction, 18, 33, 45

F

F_2, 20, 21
Fabri – Perreault interferometer, 28
fabrication, 2
Fermi level, 7, 8, 9
films, 14, 91
fixed-nuclei approximation, 24
formula, 68
fragments, 42
friction, 2

G

graphite, 2, 3, 39, 40, 46
grazing, 64, 65, 66, 67, 68, 69, 71, 73, 74, 77, 78, 80, 82, 83, 84, 85, 88, 90

H

hardness, 2
height, 81
hopping time, 29
hybridization, vii, 3, 4, 37
hypothesis, 1

I

IFM, 93
impurities, 44
incidence, 64, 65, 66, 67, 68, 73, 74, 77, 78, 80, 84, 87, 88, 90, 91
incoherent waves, 29
inelastic threshold, 32, 34, 41
inequality, 30
initial state, 63

Index

integration, 70
interdependence, 27
interface, 2
interference, 21, 22, 28, 29, 32, 34, 47, 49
inversion, 28, 29
ionicity, 2
ionization, vii, 23, 28, 32, 33, 34, 40, 43, 76
ions, 3
isotropic media, 51, 54

K

kinematic approximation, 25, 26
Kramers – Kronig dispersion relation, 2

L

linear dependence, 42
linearity, 42
local anisotropy, 41
localization, 1, 4, 5, 17, 28, 29, 30, 42, 43, 44, 47, 48, 49
lubricants, 2
luminescence, 14
Luo, 95
lying, 32, 41, 42, 43, 46, 47

M

manufacturing, 15
matrix, 22, 23, 24, 52, 62
media, 1, 51, 54
melting, 2
melting temperature, 2
metals, 2
model system, 18, 19
modeling, 44
modulating function, 21
modulus, 69
molecular clusters, 29, 32
molecular simulation, 44, 48
molecules, 18, 20, 21, 28, 29, 32, 40, 41, 43, 48, 75
momentum, 5, 22

Moscow, 96
multielectron excitations, 22, 43, 76

N

N_2, 18, 20, 21, 28, 29
nanometers, 15
Ne_2, 20, 21
nitrogen, vii, 2, 3, 5, 6, 7, 8, 9, 10, 11, 12, 13, 15, 18, 41, 47, 48, 56, 59, 76
nitrogen compounds, 2
NO, 20, 21
nonlinear optics, 100
nuclei, 24

O

O_2, 20, 21
optical potential, 31, 32
optical properties, vii, 1, 57, 83
oscillation, 75
oscillations, 26, 70
overlap, 3

P

parallel, 10, 51, 54, 55, 57, 62, 63, 64, 73, 87
permittivity, 87
phase shifts, 26
photoabsorption, 22, 62
photoelectron fluxes, 23, 29
photoelectron reflection, 21, 48
photoelectron spectroscopy, 7
photoemission, vii, 1, 25, 29, 30, 32
photons, 62
physics, 96, 98
planar compounds, 18, 38
plane waves, 11, 54
polarization, vii, 51, 52, 53, 55, 62, 63, 68, 87, 88, 90, 91
porosity, 15
probability, 28, 29, 63
probe, 1

propagation, 15, 51, 54
pseudopotential, 17, 18, 22

Q

quantum chemistry, 1
quantum mechanics, 96
quasiatomic approach, 1, 31, 32
quasiatomic origin, 19

R

radiation, vii, 1, 15, 59, 61, 62, 63, 64, 65, 66, 67, 68, 73, 74, 78, 80, 81, 83, 84, 85, 87, 88, 89, 90, 91
radius, 23, 24, 30, 45, 55, 81
recall, 69
recommendations, iv
redistribution, 9, 77
reflectivity, 14, 23, 64, 69, 85, 90
refractive index, 52, 53, 55
refractive indices, 53
relaxation, 29, 30, 31, 44, 76
resolution, 37, 43, 75, 83
roughness, 73, 82

S

scatter, 15
scattering, vii, 14, 15, 21, 23, 27, 30, 31, 32, 33, 40, 43, 47, 62, 73, 74, 75, 81, 82, 83, 84, 85, 88, 96, 99
screening, 18
semiconductors, 14
sensitivity, 15, 45
shape, 18, 20, 32, 33, 34, 39, 41, 42, 43, 45, 46, 47, 55, 59, 64, 82, 85, 88
shape resonance, 18, 20, 32, 33, 34, 39, 41, 42, 45, 47
short range order, vii, 24
simulation, 44, 48
solid state, 5
species, 31, 39, 40, 41, 42, 43, 44, 52
spectroscopy, 5, 12, 28, 32, 62

statistics, 82, 84
structural changes, 46
surface layer, 77
surface region, 83
surroundings potential, 17, 22, 23, 24, 25, 27, 37
susceptibility, 51, 62
symmetry, 3, 5, 11, 15, 17, 21, 22, 28, 29, 30, 44, 55, 63, 76, 88
synthesis, 2

T

Tchebychev polynomial series, 33
temperature, 2, 45
thermal expansion, 2
thermodynamic equilibrium, 7
thin films, 45, 91
three-dimensional model, 11
translation, 28
transmission, 1, 23, 24, 25, 26, 88
transparency, 48, 77
trapping time, 29
tunneling, 32, 40

U

united atom, 17, 19, 20
urea, 2
USSR, 93, 99, 100
UV, 14, 70

V

valence, 3, 5, 7, 9, 10, 11, 12, 14, 32, 33, 34
valleys, 18
van-der-Waals interaction, 17
variations, vii, 33, 45, 47, 48, 77
vector, 51, 53, 54, 61, 62, 64, 74, 81
velocity, 30, 52, 53, 54
vibration, 55

W

wave number, 32
wave vector, 52, 53, 54, 57
weakness, 48
wear, 2
workers, 40, 46

X

XPS, 7, 8, 9, 14
X-ray, vii, 1, 5, 6, 14, 15, 17, 18, 20, 21, 22, 23, 25, 28, 43, 44, 46, 49, 59, 62, 70, 91, 96